THE LIBRARY OF PRINTING TECHNOLOGY
MATERIALS IN PRINTING PROCESSES

Other books in the Series

DESIGN FOR PRINT PRODUCTION
H S Warford

LETTER ASSEMBLY IN PRINTING
D Wooldridge

FINISHING PROCESSES IN PRINTING
A G Martin

MACHINE PRINTING
W R Durrant, C Meacock
R E Whitworth

GRAPHIC REPRODUCTION PROCESSES
Eric Chambers

Series Editor
J E REEVE FOWKES MIOP
Head of the Department of Printing
Southampton College of Art

THE LIBRARY OF PRINTING TECHNOLOGY

Materials in Printing Processes

BY L C YOUNG, BSc FIOP FRIC

Visual Communication Books
HASTINGS HOUSE, PUBLISHERS
10 East 40th Street, New York, N.Y. 10016

Library of Congress Cataloging in Publication Data
YOUNG, LAURENCE CARVAN
  Materials in printing processes.
  (The Library of Printing Technology)
    1. Paper.   2. Printing-ink.   3. Printing machinery and supplies.   I. Title.
Z247.Y65   686'.028   71–38323
ISBN 0 8038 4666 5

Printed in Great Britain by
W & J Mackay Limited, Chatham

CONTENTS

Contents

Contents

Contents

viii

LIST OF PLATES

Editor's Preface

Printing, in common with other great industries, is caught up in nuclear-age technology. For some printers this is a painful experience: the deep roots of five centuries of craft tradition resist the pull of technology. The pride of craftsmanship, with all its ingrained mystique, conflicts with the self-assurance of technological skills – skills which reduce the art and mystery of printing to binary numbers.

Already great changes have taken place in the printing industry: more are to come and the rate of change will increase. What is happening needs to be understood by printers if they are to remain masters of their destiny. More printers must be trained as technicians and technologists if the benefits of the 'knowledge explosion' are to be fully exploited.

Education and training for the printing industry has changed to meet the challenge. Courses for technicians and technologists have been established in many countries. In Great Britain, the Institute of Printing has fostered National Certificate, Diploma and Degree courses, and the City and Guilds of London Institute offers examinations for a Printing Technician's Certificate.

These new courses need books, and just as the industry is in danger of being overtaken by events, so too is the printer's library. This danger was seen by leading educationalists, among whom were Charles L Pickering, OBE, teacher of printing and for many years HM Inspector of Schools with special interest in printing education; H M Cartwright, former Principal of Bolt Court School of Photo-Engraving and a leading authority on photo-mechanical processes; and L T Owens, Principal of the London College of Printing and a member of the Printing and Publishing Industry Training Board.

This series is unique in that it embraces, for the first time, the complete spectrum of printing technology – pre-production design and planning, letter assembly, graphic reproduction and printing surface preparation, machine printing, print finishing, and the technology of printing materials.

The principles of printing – transfer of a pigment from an image-bearing surface to a substrate – are simple. In practice, however, the processes become complex because printing plates, printing inks and paper vary from job to job to satisfy the whims of designers and the demands of print buyers. The permutations are so numerous that every job presents its own special problems, aggravated by the instability and unexpected behaviour of some substrates and inks. The patience of the printer is tested further by the interaction of adhesives with paper and ink in the print finishing workshop.

It is fundamental to the control of printing processes that the physical and chemical nature of materials should be understood. This book has been written primarily for the printing technician whose job it is to control successfully the processes of printing surface preparation, machine printing and finishing; but it will appeal also to craftsmen who not infrequently encounter the frustrations of unexpected failures caused by the interaction of materials and who for their own satisfaction want to know why.

J E Reeve Fowkes

Acknowledgments

I would like to record my thanks to the many people in industry and education who have helped me in writing this book; to my colleagues on the staff of Watford College of Technology and in particular to V J Turner for his helpful advice on the chapters dealing with printing ink and for his assistance with the specimen ink formulations; to T Pearman for his valuable comments on the chapter on bookbinding materials; to G H Atkins and J W Archer for their advice in the preparation of the line diagrams; to J Graham the college librarian, and N Moore the college printing information officer; and to B H Davey, P E Watkins and T J Cowley for their general advice and encouragement.

I would also like to thank R J Pierce of the Fishburn Printing Ink Co. and S M White of the International Publishing Corporation for reading and commenting on the chapters dealing with printing ink and paper; E W Peacock of PIRA for proving to be a mine of information for the appendix on testing instruments; and my former colleague, F Pateman, for his general advice.

Grateful thanks are also extended to the following persons and organisations for supplying and giving permission to reproduce the figures, plates and text matter as listed below:

The British Federation of Master Printers. Extract from 'Metrication—A Programme For Change'

The British Paper and Board Makers Association, Figs. 5.1 and 5.3

P J Bryant and the Institute of Printing for Figs. 7.3 to 7.10 which first appeared in *Printing Technology* Vol. 12, No. 1, April 1968

J Chatillon & Sons Ltd. Plate 16

Mickle Laboratory Engineering Co. Plate 18

B Clark, Reed Engineering and Development Services Ltd. Plates 1–6

Fry's Metals Ltd. Figs. 3.5, 3.6, 3.7, together with some associated text from their publication *Printing Metals*

H E Messmer Ltd. Plates 7, 8, 9, 10, 11, 13, 14, 15

J R Parker, Engineering and Scientific Service Division, Bowaters Co. Ltd. Fig. 9.20

Acknowledgments

PIRA. Plates 12, 17, 19, 20, Figs. 9.1, 9.24
Torrance & Sons Ltd. Figs. 13.1 and 13.6
Carl Zeiss & Co. Ltd. Fig. 9.12
Sir Isaac Pitman Ltd. and F Pateman for permission to include a number of short extracts from the textbook *Printing Science* by F Pateman and L C Young
British Standards Institution and the Technical Association for the Pulp and Paper Industry. Lists of standards in Appendix C.

Finally a very special word of thanks to my wife Elizabeth, and to Alison, Barbara and Simon for their patience and understanding while this book was being written.

Laurie Young

1. Introduction – the nature of materials

The average man, whoever he may be, might quite reasonably imagine that the materials of the printing industry must be simply two – paper and ink, because everyone from the ancient Chinese onwards has understood the equation

$$\text{Paper} + \text{Ink} = \text{Printing}$$

The average printer soon finds out that paper and ink are two words which cover a tremendous variety of products – each one behaving differently in the printing process. He also discovers that he has to work with many other materials, including metals, alloys, plastics, photographic materials, leather, cloths and adhesives. In bringing a study of all these materials into a single volume for the first time, this book is an attempt to give those involved with printing a better understanding of the behaviour of materials in the various processes of the industry.

On the face of it these printing materials – papers, printing inks, plastics, metals, adhesives, etc, are completely unrelated to each other, but when we have a closer look at their basic construction, we find the same relationship that exists between all materials in the *structure of matter*.

ELEMENTS AND ATOMS

All the materials we shall be considering are built up from the 104 *elements* which form the building units of our world. We can define an element as a substance which cannot be split up into anything simpler by chemical means. Some printing materials, in particular the platemaking metals like zinc and copper, are elements, although they are not used in a completely pure form. If we were to cut a copper plate into two and were able to go on dividing it into smaller and smaller parts, we would eventually reach the smallest possible piece of the metal, an *atom* of copper. Since there are about 10 000 000 000 000 000 000 000 atoms in every gramme of copper, we would obviously not be able to make this division, nor would we be able to see a single atom under the microscope. Since the smallest whole unit of an element is an atom, there are 104 basically different types of atoms.

THE STRUCTURE OF ATOMS

If we go one stage further we find that these 104 different atoms are all built up from the same particles of matter, the most important being *electrons, protons* and *neutrons*. A copper atom differs from an oxygen atom or a carbon atom in the number of these particles that it possesses and in the way in which they are arranged. Whilst the proton and the neutron have almost the same mass, the mass of an electron is only 1/1836 of that amount and hence its contribution to the mass of the atom is negligible. Protons and electrons are electrically charged, each proton carrying a unit positive charge and each electron a unit negative charge. Neutrons, as the name suggests, are electrically neutral. Since atoms are themselves neutral, the number of electrons must always be the same as the number of protons.

Atoms can be pictured as very small solar systems, in which the negatively charged electrons move in orbits around a positively charged heavy central nucleus. Taking one example, an atom of aluminium consists of a central nucleus containing 13 protons and 14 neutrons, with 13 electrons orbiting round that nucleus.

COMPOUNDS AND MIXTURES

In practice, of course, very few of our everyday materials are elements. Many are *compounds* consisting of atoms of two or more elements linked together as *molecules*. Many of the materials considered in this book are compounds – substances like nitric acid, polythene and toluene. An even greater number of materials are *mixtures*, containing variable quantities of two or more substances. Printing inks, varnishes, adhesives, papers and air are all mixtures.

a.   Sodium 2.8.1   b.   Chlorine 2.8.7

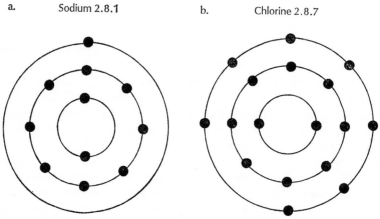

1.1 Arrangement of electrons in atoms of sodium and chlorine.

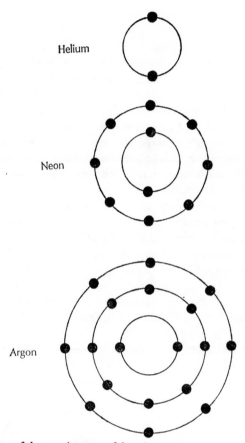

Helium

Neon

Argon

1.2 Arrangement of electrons in atoms of three *inert gases*.

LINKAGES BETWEEN ATOMS

The orbiting electrons of an atom are arranged in shells at different distances from the central core, rather like the layers of an onion. Each shell can only hold a certain number of electrons. For example, an atom of sodium has 11 electrons, with 2 of these in a completed inner shell, 8 in a completed second shell and one odd electron in an incomplete third shell. This is shown diagramatically (fig. 1.1a). The 17 electrons in an atom of chlorine fill the first two shells and have a third shell containing 7 electrons, one less than a full complement (fig. 1.1b). The elements whose atoms have completed electron shells form a unique family of exceptionally stable elements – the so-called *inert gases* (fig. 1.2). *When two or more atoms combine together they do so in order to*

3

achieve this extremely stable inert gas structure of complete shells. This is done in two main ways.

In the formation of *ionic bonds*, one or more electrons are transferred from one atom to another in such a way that both atoms achieve the inert gas structure. For example, sodium combines with chlorine by giving its odd third shell electron to the chlorine atom which is then able to complete its own third shell (fig. 1.3). Since electrons are negatively charged, a molecule formed by an electron transfer consists of two parts – the atom which has given up an electron becoming positively charged and the atom which has gained an electron becoming negatively charged. These charged atoms are known as *ions*. The strength of ionic bonds is due to the force of electrostatic attraction between the positive and negative ions.

Elements differ greatly in their power of attraction for electrons. For example, chlorine atoms needing only one electron to achieve an inert gas structure are strongly electron-attracting or *electronegative*. In contrast, an element like sodium with one odd outer electron to dispose of has no affinity for electrons and is strongly *electropositive*. Clearly, ionic bonds are formed most readily between electropositive and electronegative elements. Many inorganic compounds including metal salts like copper sulphate, lead chromate or ferric chloride and acids like nitric and hydrochloric are formed by ionic linkage.

A second way in which atoms can combine together to achieve the stable

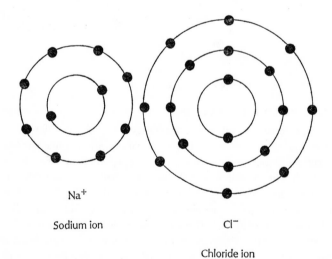

Na$^+$

Sodium ion   Cl$^-$

                              Chloride ion

1.3 Arrangement of electrons in a molecule of sodium chloride.

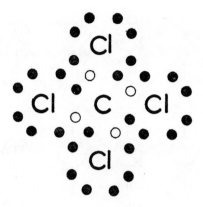

only outer electrons shown

○ represents electrons from carbon

● represents electrons from chlorine

1.4 Arrangement of outer electrons in a molecule of carbon tetrachloride.

electronic arrangement of the inert gases is by the sharing of electrons between the combining atoms (fig. 1.4). These *covalent bonds* are often formed by atoms of elements which do not show strong electropositive or electronegative character. Carbon is one such element and *organic chemistry* is concerned with the hundreds of thousands of carbon-containing chemicals formed primarily by these covalent linkages.

Compounds formed by these two different methods of linkage – ionic and covalent – each have a distinctive character. Ionic compounds owe their stability to the strength of the electrostatic attraction between oppositely charged ions, and we find that most metal oxides and salts, and inorganic compounds will withstand high temperatures. On the other hand covalent bonds are much more easily broken, and many organic compounds burn in air at comparatively low temperatures, some of these compounds providing us with our fuels in such mixtures as coal, natural and coal gas, petrols, paraffins and oils.

As well as being attracted to each other the positive and negative ions in an inorganic compound like sodium chloride are also attracted to neighbouring ions carrying an opposite charge (fig. 1.5). This force of attraction between neighbouring molecules is reflected in the very high melting and boiling points

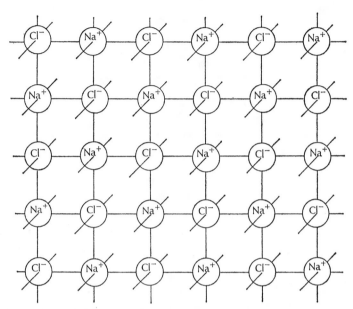

1.5 Sodium and chloride ions arranged in a cubic lattice structure in sodium chloride crystals.

of ionic compounds, since these temperatures are a measure of the energy which has to be supplied to give individual molecules the partial freedom of the liquid state or the complete freedom of the gaseous state. The forces of attraction between molecules of covalent compounds are much weaker, and as a result organic chemicals have much lower melting and boiling points than inorganic molecules of a similar size. In fact many simple organic substances are liquids or even gases at room temperature.

Another characteristic of ionic compounds is that they are capable of conducting electricity when in aqueous solution or when in a fused state. Since the ions themselves act as the carriers of the electrons through the liquid, covalent compounds are not able to behave in this way.

DIPOLE MOMENTS

So far we have only considered bonds formed by electron donation or electron sharing, but this is not the whole story. As one so often finds, in addition to black and white there are several shades of grey. When two identical atoms form a covalent bond, as in a hydrogen molecule, H_2, the bond can be formed on the basis of precisely equal sharing of the two electrons (fig. 1.6). However, if a covalent bond is formed between atoms of two elements having different

6

electronegativity, the atom which attracts electrons more strongly will get rather more than its fair share. In other words covalent bonds can have a partially ionic or *polar* character, and molecules formed in this way are said to have *dipole moments*, *ie* the centres of the positive and negative charges are separated by a certain distance. In effect, this means that these molecules have positively charged and negatively charged parts. For example the unequal sharing of electrons between atoms of oxygen and hydrogen in molecules of water, give it a strong dipole moment and polar character (fig. 1.7).

As one might expect, the polar character of a bond is reflected in the properties of the molecule of which it forms a part. The force of attraction between neighbouring molecules with a polar character is greater than for non-polar molecules (fig. 1.8), and so melting and boiling points are higher than those of non-polar molecules of similar molecular size (Table 1.1). The polar character of a substance also affects its ability to dissolve other substances or to be dissolved itself. In general one finds that 'like dissolves like' *ie* polar solvents dissolve polar solutes and non-polar solvents dissolve non-polar solutes. These ideas are further developed on page 172.

GASES, LIQUIDS AND SOLIDS

If a small piece of zinc is heated in a nickel crucible over a Bunsen burner it

1.6 Equal sharing of electrons in a molecule of hydrogen.

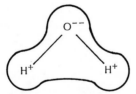

1.7 Unequal sharing of electrons in a molecule of water.

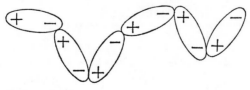

1.8 Association of polar molecules.

7

will soon reach the temperature (430°C) where the zinc becomes a liquid. If the heating continues, it can reach a temperature of 927°C when the liquid zinc changes into a vapour or gas. Soon after it escapes from the surface of the metal the zinc gas reacts with oxygen in the air to form 'cobwebs' of white zinc oxide, so we discover rather unexpectedly that zinc, which we normally think of as a very solid solid, can also be a liquid or a gas.

Carbon dioxide, the gas that we breathe out, can be converted into a solid at low temperatures. Ice, a solid, changes into a liquid (water) above 0°C and into a gas (steam) above 100°C. In fact, all substances can exist as solids, liquids, and gases at particular temperatures and pressures. It should be remembered that these changes of state make no difference to the substance chemically. Heating iron and sulphur leads to a chemical change, but heating zinc to change it from a solid to a liquid is only a physical change. A chemical change involves the forming or the breaking of chemical bonds between atoms.

In all solid substances, the atoms, ions or molecules are closely packed together. They may be arranged in a regular three-dimensional structure (crystalline solids) or alternatively they may be in a completely random arrangement (amorphous solids). Although held by its neighbours, each atom is in a constant state of vibration. The amount of vibration is increased as heat is applied to the substance.

Table 1.1   Relationship between dipole moments and boiling points for molecules of similar size

Compound	Structure	Dipole moment	Boiling point
Methane	H | H—C—H | H	Nil	−162°C
Chloroform	H | H—C—Cl | H	Considerable	−24°C
Nitromethane	H | H—C—NO_2 | H	Very large	+101°C

At a certain temperature (the melting point) all the atoms of a substance are moving so fast that the force of attraction between neighbours is not strong enough to prevent them moving about. This is the liquid state, where atoms no longer have to stay in fixed positions but are able to slide over their next door atoms to new neighbours. Whilst the atoms in a liquid have this increased freedom they still tend to cling together as a whole. This can be demonstrated well by placing a drop of mercury on a plate, or a drop of water on a greasy surface.

The speed of movement of the atoms in a liquid goes on increasing as the temperature is raised until a point is reached where all the atoms have sufficient energy to overcome the force of attraction that they have for one another. Rather like rockets being fired clear of the earth's gravitational pull, the atoms are now going fast enough to escape from the surface of the liquid. The atoms of a gas move freely in all directions and when enclosed they occupy all the available space. In this explanation we have talked about the motion of atoms in an element, but precisely the same reasoning applies to the movement of molecules in compounds, most of which can exist either as solids, liquids or gases. Water is familiar to us in each of the three states of matter. In ice, the molecules are held in a crystal structure, without the freedom to change places with one another. When the temperature of ice is raised above O°C to become water, the molecules, moving faster, have gained this freedom, although they still tend to cling together as a whole. At 100°C when the water becomes steam, the molecules are moving sufficiently fast to escape and move into any available space.

SMALL MOLECULES AND LARGE MOLECULES

Papers, printing inks, and most other printing materials are mixtures of a great variety of substances. These substances fall into two main classes of compounds. Firstly, there are the *inorganic compounds*, like the silver salts used in photography, ferric chloride and nitric acid in etching, coloured inorganic pigments like lead chromate in printing inks, white pigments like titanium dioxide in papers and printing inks, and the dichromate salts used in plate-making. Secondly there is the much larger group of *organic compounds*, all containing carbon and including cellulose, drying oils, resins, plastics, solvents, diazo compounds and organic pigments.

In general the inorganic compounds have small molecules with relatively few atoms, eg $AgBr$, $K_2Cr_2O_7$, TiO_2, $FeCl_3$. The linkages between the atoms are mainly ionic and as one might expect the compounds are, on the whole, stable, and have high melting points. The organic compounds show a much greater variation in size, ranging from the simple molecules of solvents like

9

ethyl alcohol, C_2H_6O, to the giant molecules of natural polymers such as cellulose or synthetic polymers like polyvinyl acetate. To give some idea of the size of these large molecules, one molecule of cellulose may contain as many as 40,000 atoms of carbon, hydrogen and oxygen. For these organic compounds, there is a broad relationship between a substance's state at ordinary temperatures and its molecular size, and as a general rule one finds that the very small molecules are gases, *eg* methane CH_4, ethane C_2H_6, ethylene C_2H_4; slightly larger molecules are liquids, *eg* ethyl alcohol C_2H_6O, glycerol $C_3H_9O_3$; and the large molecules and polymers are solids. As we saw earlier (fig. 8) when an organic molecule has some polar character, its melting point and boiling point will be higher than that of a purely covalent compound of similar molecular size.

SOLUTIONS AND DISPERSIONS

A solution may be defined as a very fine uniform mixture of two or more substances. We refer to the dissolved substance as the *solute* and the second substance, which does the dissolving, as the *solvent*. When common salt dissolves in water, the dissolved particles exist as single molecules of sodium chloride or as single sodium and chloride ions – and the mixture is known as a *true solution*. When gelatin is dissolved in warm water the dissolved particles are much larger than the molecules or ions in a true solution although they are still too small to be visible under an ordinary microscope. This type of solution is correctly known as a *colloidal dispersion*, and the term colloid or colloidal substance is used to describe those materials which can be dispersed as particles ranging in size from about 1 nanometre (nm) to 1 micrometre (μm) diameter. When the size of dispersed particles exceeds colloidal dimensions the mixture becomes a coarse *suspension* or dispersion. When one liquid is dispersed in a second liquid, with which it will not mix, the result is an *emulsion*.

The difference between true solutions and other types of dispersion can be very simply demonstrated by shaking small quantities of a number of common materials with water in a test tube. Substances like common salt or washing soda will dissolve easily to form a clear true solution, containing minute particles that are invisible even under a microscope. If we shake up a mixture of sand and water the result is a suspension in which the particles are relatively large and visible. This particular suspension is unstable and the sand will soon settle out, but it is possible for suspensions to be prepared which are more stable.

When a small amount of soil is shaken with water in a test tube a number of things will happen. Some of the materials in the soil will immediately dissolve and form a true solution. Another part will be suspended in the water, but the large particles will soon settle out again. However, when the settling is complete, the liquid will still appear cloudy. This cloudiness is due to the effect of particles which are intermediate in size between those which have dissolved and those which have been temporarily suspended. These particles in colloidal dispersion are too small to be seen by the eye but they are large enough to scatter rays of light and so cause cloudiness.

If some methylated spirit is shaken with water the two liquids will easily mix together. On the other hand paraffin and water form quite separate layers. If they are vigorously shaken together, one of the liquids becomes dispersed in the other in the form of fine droplets. A dispersion of this type is called an emulsion. An emulsion of paraffin and water is unstable, the two liquids soon returning to their separate layers. More stable emulsions like milk, white glues, and emulsion paints are familiar materials in our everyday world.

Printing materials provide many examples of these different types of solution and dispersion. Etchants like nitric acid or ferric chloride, dichromate salts used in sensitising polymers, sodium thiosulphate (hypo) used in fixing photographic films, copper salts used in plating baths, are all compounds which dissolve in water to form true solutions. When gelatin or gum arabic are dissolved in water, and when natural or synthetic resins are dissolved in organic solvents, the resulting mixtures are colloidal dispersions. Printing ink normally consists of a suspension of insoluble particles of pigment dispersed in a liquid vehicle. During a litho printing run some water is carried into the ink to form a water-in-oil emulsion and it is possible for small quantities of the ink to work their way back into the fountain solution as an oil-in-water emulsion.

2. Metals for platemaking

Metals are as essential to the printing industry as they are to any other highly mechanised modern industry. In addition to their obvious use in the construction of machines, their importance to the printer rests on the fact that most printing ink is applied from a metal surface and that in some cases printing ink is applied to a metal surface. In other words, metals are materials that we 'print from' *eg* copper or zinc plates, type alloys, and materials that we 'print on', *eg* aluminium foil, tin plate. Normally printing surfaces are either prepared by casting alloys of lead, tin and antimony into the required shape or by the photochemical treatment of the surface of a metal plate or cylinder. A thin layer of a very hard metal may be finally electrodeposited on to a printing surface to improve its wearing quality. In this chapter we will be considering the properties of certain metals in relation to their uses in platemaking. Alloys for type casting are dealt with in the next chapter.

Although there are about eighty metals, only a few of them have the properties which are required for a printing plate. Obviously, these properties must depend on how the plate is going to be processed, on the nature of the printing process and on the length of run required, but in general terms a metal is needed which is malleable, with good wear resistance and strength, a suitably fine grain structure and reasonable cost. Other essential properties may include good stability to heat and the ability to be easily etched. Four metals which have proved themselves in platemaking are copper, aluminium, zinc and magnesium. Thin layers of two other metals, nickel and chromium, are sometimes electrodeposited on to a printing plate to improve its wear resistance. Before considering these metals individually we will first make some broad comparisons between them, based on some of their more important properties.

Table 2.1 Specific gravities

	Copper	Aluminium	Zinc	Magnesium	Chromium	Nickel	Lead
Specific gravity	8·9	2·7	7·1	1·7	7·1	8·9	11·4

The wide range of specific gravities shown in Table 1 is striking. At one extreme there is the high specific gravity of lead (11·4) which is largely responsible for the great weight of letterpress formes and for subsequent handling problems; at the other, aluminium and magnesium are among the lightest of metals and the ease of handling of a thin aluminium plate has been one of the advantages which has assisted the growth of the litho process. It should be noted that the specific gravity of a metal can be increased slightly by a mechanical treatment such as rolling or hammering, eg the specific gravity of aluminium may vary from 2·58 to 2·69.

Table 2.2 Melting points

	Copper	Aluminium	Zinc	Magnesium	Chromium	Nickel	Lead
Melting point °C	1083	660	419	650	1875	1455	327

Lead, the major constituent of all type alloys, has a low melting point, which allows it to be easily cast. However the low melting point of zinc and the fact that it is not dimensionally stable above 200°C is far from being an advantage because letterpress blockmaking includes the process of 'burning-in' the exposed coating in order to convert it into a hard etchant-resistance stencil. Apart from dimensional changes, it has been shown that when magnesium and zinc are processed at 300°C their hardness is considerably reduced, a point which we will return to in the next section of this chapter.

Table 2.3 Hardness

	Copper	Aluminium	Zinc	Magnesium	Chromium	Nickel	Lead
Brinell Hardness	35	24	31	33	100	100	4–6

It is reasonable to suppose that the wear resistance of a printing surface is related to its hardness and that this property is more important for the metals used in making letterpress blocks than for the aluminium used for offset litho plates.

At one time it was generally believed that the hardness of letterpress blocks increased greatly during a press run. An early experiment carried out by PATRA (now PIRA) showed that these increases in hardness are quite small, particularly when they are compared with decreases liable to occur in the blockmaking process. The result of this investigation, carried out on a

Miehle using a machine coated paper and black ink, are summarised in Table 2.4.

Table 2.4 Hardness changes with block-making and printing

Operation		% Hardness Change		
		Copper	Magnesium	Zinc
(A)	Blockmaking	−2·6	−23·2	−5·3
(B)	89000 Impressions	−2·3	+0·6	+0·8
(C)	Further 61000 impressions with increased pressure	+3·3	+1·4	+0·3
	Net effect of (B) + (C)	+0·9	+2·0	+1·1

It was also shown that when magnesium was processed at 300°C its hardness fell by 15% and that of zinc by 12%. The hardness of copper was not seriously affected until 400°C was exceeded. The report stresses the need for some form of temperature control in the 'burning-in' operation and concludes that the maximum safe temperatures are 400°C for copper, 250°C for magnesium and 200°C for zinc.

The mechanical properties of any metal, including its hardness and tensile strength, depend to a great extent on its purity and physical condition. Generally speaking the more a metal is purified the softer it becomes. Properties can often be improved by blending with another metal or non-metal in an alloy. For example zinc alloys are now available for blockmaking with a much finer crystalline structure than the pure metal. The physical condition of a metal depends on whether it is 'as cast' or whether it has been worked in some way. For example, the hardness of copper is greatly increased by rolling.

THE LITHOGRAPHIC PROPERTIES OF METALS

A litho plate is designed so that its image areas will welcome greasy ink and repel water, whilst its non-image areas welcome water but repel ink. A so-called 'deep etch' litho plate carries an image consisting of a hard polymeric lacquer covered with a greasy ink and a non-image consisting of grained aluminium covered with a thin film of water (which may contain acidulated gum). For very long press runs use is made of bimetallic litho plates, with one metal forming the image area and a different metal the non-image (fig. 2.1). These plates depend on the fact that neither water nor oil wet all metals with the same ease. The metal chosen for the image areas must be oleophilic or oil-loving in character, whilst the non-image metal must be hydrophilic or water-loving.

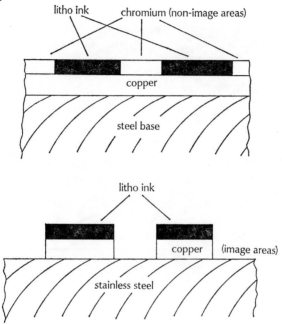

2.1 Multimetal litho plates (a) Copper image/Stainless steel non-image. (b) Copper image/Chromium non-image.

The relative wettability of metals in the lithographic process has been investigated by the measurement of *contact angles*. When a drop of liquid is in contact with a solid surface it may wet the surface and spread completely over it or it may break up into droplets having a particular shape and contact angle (fig. 2.2). This contact angle is a measure of the liquid's ability to wet the surface. Because the process of lithography involves two immiscible liquids, the contact angle between one liquid and the metal must be measured in the presence of the second liquid. These interfacial contact angles may be measured by introducing a small drop of oil from a pipette under the metal plate immersed in water, and projecting the image of this oil drop on to a screen so that its contact angle can be measured (fig. 2.3). In the original work

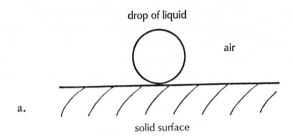

drop of liquid

air

a.

solid surface

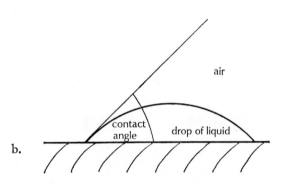

air

contact
angle

drop of liquid

b.

liquid spread

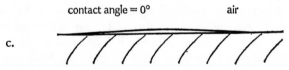

contact angle = 0°

air

c.

2.2 Degrees of wetting of a solid surface by a liquid:

      a.   No wetting
      b.   Partial wetting
      c.   Complete wetting

at PIRA, R A C Adams measured those interfacial contact angles on various metals for an oil, consisting of a 5% solution of oleic acid in liquid paraffin. This oil mixture was chosen because it has similar surface chemical properties to a basic lithographic varnish. The results, shown in Table 2.5, provide a *lithographic series* placing the metals in order of their oleophilic character, zinc with an interfacial contact angle of 30° being the most oil-loving.

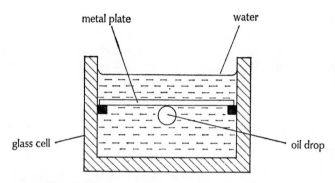

2.3 Measurement of interfacial contact angles.

Theoretically one would expect a bimetallic plate to be made up of metals which are well separated in the series, an oleophilic image near the top and a hydrophilic non-image near the bottom. This is borne out in practice by such combinations as copper/chromium and copper/stainless steel (fig. 2.1).

Table 2.5 Lithographic series of metals

	Interfacial Contact Angle	
Metal	*Metal with oxide film*	*Metal after removal of oxide film*
Zinc	30°	30°
Silver	64°	50°
Copper	77°	60°
Brass	86°	75°
Nickel	100°	83°
Stainless steel	110°	86°
Aluminium	140°	50°
Chromium	150°	107°

The lithographic properties of a metal can be greatly modified by the chemical treatment of its surface. For example, although aluminium is low in the lithographic series and has hydrophilic properties which make it a most effective non-image metal, the removal of the layer of oxide from its surface reduces its contact angle to 50° and renders it relatively oleophilic.

COPPER

Copper is unique in the printing industry because it is used in the production of printing surfaces for letterpress, lithography and gravure. It provides fine line and halftone blocks for the letterpress process, plates and cylinders for gravure and it may be a component in bimetallic plates for lithography. Its printing applications do not end there, since it is also used in the production of electrotypes for duplicate letterpress plates and as a thin coating on steel inking rollers on presses.

The most outstanding property of copper is its high conductivity of heat and electricity, this being second only to that of silver. More than half of the world's production of copper is made into wire to carry electricity. Copper can be deposited electrolytically in thick or thin layers on to the surface of another metal. Gravure cylinders often consist of a skin of copper laid down in this way on to a steel core. The physical condition of an electrodeposited coating may be varied by adjusting the conditions such as the bath concentration, the current density and the temperature.

Copper's malleability and ductility make it possible for foil less than $50\,\mu m$ thick and wire $25\,\mu m$ in diameter to be commercially produced. The metal has a fine crystalline structure so that chemical etching or mechanical engraving can produce very small relief dots for letterpress or very fine recessed cells for gravure. Although it is only moderately hard (Table 2.3), its actual press life will depend to a large extent on the operating conditions. If necessary, wear resistance can be improved by putting on a facing of a harder metal such as chromium.

We have already seen that copper stands high in the lithographic series of metals and its oleophilic character makes it suitable to form the image areas of bimetallic plates (fig. 2.1). The affinity of copper for ink has been exploited in other ways in the printing industry. The durability and affinity for ink of a deep etch image can be improved by the application of a thin layer of copper and in the same way steel ink rollers can be made more ink receptive.

Copper is extremely resistant to atmospheric corrosion, chiefly as a result of the formation of a film of oxide which unlike rust on ferrous metals does not absorb moisture. On long exposure to an atmosphere containing carbon dioxide or other acid vapours, the oxide skin is converted to sulphides or to

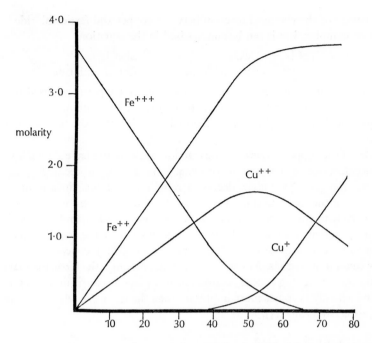

total amount of copper dissolved (% of theoretical exhaustion)

2.4 Changes in ionic concentrations in the etching of copper (Saubestre).

complex basic sulphates and carbonates in a green coating, such as can be seen on the copper roofs of churches and other buildings.

Concentrated sulphuric and concentrated hydrochloric acids have little effect on copper but it is attacked by nitric acid. Ferric chloride is the normal etchant for copper in the production of letterpress and gravure printing surfaces. Although a salt, ferric chloride is strongly hydrolysed to give extremely acidic solutions.

$$FeCl_3 + 3H_2O \rightleftharpoons Fe(OH)_3 \qquad + \qquad 3HCl$$
$$\text{ferric hydroxide} \qquad \text{hydrochloric acid}$$
$$\text{(weak base)} \qquad \text{(strong acid)}$$

The solutions used in the etching of copper all appear to have pH's below zero. One advantage of ferric chloride as an etchant is that no gas is produced in the reaction. If gas is produced, small bubbles held on the surface of a metal, perhaps in the cells of a gravure plate could locally prevent contact between the etchant and the metal and so slow down the rate of etching in that area.

The nature of the chemical reaction between copper and ferric chloride is somewhat complex, but it can be summarised in the equation

$$Cu + 2FeCl_3 = \quad CuCl_2 \quad + \quad 2FeCl_2$$
$$\text{cupric chloride} \qquad \text{ferrous chloride}$$

This is an oxidation/reduction reaction in which the copper is oxidised first to the cuprous (Cu^+) and then cupric (Cu^{++}) state. In bringing about the oxidation the ferric ions (Fe^{+++}) are themselves reduced to the ferrous state (Fe^{++}).

Studies of the copper/ferric chloride reaction by Saubestre have shown how the concentrations of the various ions change as etching proceeds (fig. 2.4). The ferric ion (Fe^{+++}) content decreases rapidly and almost linearly up to about 50% of the theoretical exhaustion of the bath. The ferrous ion (Fe^{++}) rises inversely to the decrease in the ferric ion concentration, and the cupric ion (Cu^{++}) content rises until about 50% exhaustion of the bath has been reached and then falls slowly. The cuprous ion (Cu^+) concentration is negligible until the bath is over 50% exhausted. Although the ferric ion plays the main role in etching copper, cupric ion also plays a part from the beginning, and actually becomes more important than the ferric ion when the bath s 35% exhausted (fig. 2.5).

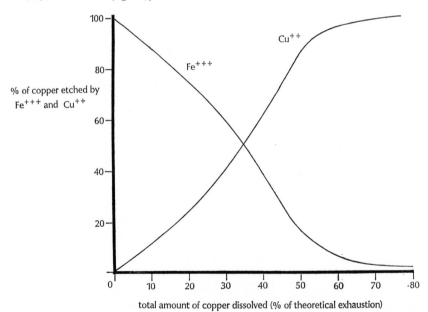

2.5 The roles of ferric and cupric ions in the etching of copper (Saubestre).

ZINC

Zinc is the metal normally used to produce letterpress line blocks. In this country copper is generally preferred for halftone blocks but elsewhere these too are often made from zinc. In the past zinc was an important litho plate metal, but a very large proportion of today's plates are made of aluminium.

We have already seen that its low melting point and poor dimensional stability above 200°C make zinc a difficult material to process. At ordinary temperatures it is brittle and crystalline but between 120°C and 150°C it becomes pliable and can be rolled into sheets. When heated over 200°C it is actually more brittle than when it is cold. A study of the surface of zinc and copper under the microscope shows that zinc has a much coarser crystalline structure, and this is the main reason why copper is usually preferred for fine line or halftone work, despite its higher cost. In recent years special zinc alloys have been developed for powderless etching. These 'microzincs' have a much finer crystalline structure than ordinary zinc and they will etch almost as quickly as magnesium.

Zinc has similar strength to aluminium but it is less elastic. If equal strips of aluminium and zinc are suspended and loaded with equal weights, both metals will stretch but when the weights are removed the aluminium recovers almost completely whilst the zinc does not. On long litho runs 750mm × 1000mm zinc plates have been shown to stretch 3–6mm round the cylinder. With only a small amount of water vapour present in the air there is only a very slight reaction between zinc and oxygen. This reaction is greatly increased at higher humidities and if a zinc plate is left with a film of water on its surface, oxidation will proceed far more rapidly than if it were dry. The layer of oxide formed on the surface does provide some protection from further attack, providing that the atmosphere is free of acid vapours. Unfortunately an industrial atmosphere is not free of these vapours (over 5 000 000 tons of sulphur dioxide are discharged over Britain each year), and under these conditions zinc is steadily corroded. Zinc blocks and plates should be dried and stored in a room with a low relative humidity or kept in a moisture-proof package.

Since zinc is electronegative to iron, it protects ferrous metals sacrificially, the slowly corroding zinc being attacked leaving the iron undamaged. The corrosion resistance of iron or steel is therefore improved by the application of a protective coating of zinc, the process known as 'galvanising'. The coating may be applied by simply dipping the article into a bath of molten zinc or alternatively the metal may be electrodeposited.

Zinc is readily attacked by dilute hydrochloric, sulphuric and nitric acids. The purer the metal, the slower will be the reaction. Hydrogen gas is given off in the reactions with hydrochloric or sulphuric, but not with nitric. The course of the reaction with nitric acid depends on the strength of the acid. In etching zinc blocks the solution is usually about a 10% solution of concentrated nitric acid in water and the products of this reaction are mainly zinc nitrate and nitric oxide gas.

$$3Zn + 8HNO_3 = 3Zn(NO_3)_2 + 2NO + 4H_2O$$

If the solution is less concentrated, no gas is evolved and ammonium nitrate and zinc nitrate are formed.

$$3Zn + 8HNO_3 = 3Zn(NO_3)_2 + 4H_2O + 2NO$$

If concentrated nitric acid is used then the brown fumes of nitrogen peroxide are given off and the main reaction is given by the equation

$$Zn + 4HNO_3 = Zn(NO_3)_2 + 2H_2O + 2NO_2$$

The very slight etch into a 'deep etch' litho plate may be made with a mixed solution containing calcium chloride, zinc chloride, ferric chloride and hydrochloric acid. Both ferric chloride and hydrochloric acid attack zinc, according to the following equations.

$$Zn + 2FeCl_3 = ZnCl_2 + 2FeCl_2$$

$$Zn + 2HCl = ZnCl_2 + H_2$$

ALUMINIUM

Aluminium can be regarded as one of the key materials in the modern printing industry for a number of reasons. The great majority of plates for litho, the fastest growing printing process, are made of aluminium. The lightness and strength of aluminium alloys make them common components in a great deal of printing equipment including composing room furniture. Aluminium foil is a very important packaging material.

Although 100 years ago aluminium was a precious metal worth £7 an ounce, today the world output of the metal ranks next to that of iron and steel. This swift rise to importance is largely due to its remarkable properties. It has a very low specific gravity compared with other metals (see Table 2.1) and, although pure aluminium is not particularly strong, its alloys have great strength. It is ductile, malleable (eg foil 50μm thick), fairly soft, and weight for weight is a better conductor of electricity than copper.

Aluminium can be rolled into thin flexible sheets for litho plates and, since it is a lighter colour than zinc, it is easier to see an image on its surface. Both aluminium and zinc litho plates are grained to improve their capacity to hold water and to hold an image, a process which more than doubles the surface area of the plate. Of the two metals aluminium is capable of taking a finer grain than zinc. One of aluminium's outstanding advantages is its ability to protect itself against corrosion by the atmosphere. It is an active metal and a film of aluminium oxide quickly forms on its surface in air. Fortunately, although this oxide layer is thin, it is hard and well 'keyed' on to the metal, and is able to protect the reactive aluminium underneath from further attack. Incidentally, it also makes the soldering of aluminium difficult.

This film of oxide can be thickened up to 15μm by the process known as 'anodising'. The metal is suspended in a bath containing either chromic or sulphuric acid solutions, and an electric current is passed as in electroplating. Oxygen is steadily released from the solution, reacts with the aluminium and builds up a layer of aluminium oxide on the surface. By varying the reaction conditions the oxide layer can be laid down either as a soft grey coating or as a hard bright transparent film. Litho plates with improved wear and corrosion resistance are produced from anodised aluminium.

We have already seen that while copper is hydrophobic, aluminium is hydrophilic. The fact that aluminium has a greater affinity for water and thus a lower affinity for ink than zinc is important in litho. Other things being equal, aluminium will run cleaner than zinc or, to put it another way, aluminium can be run with less water than zinc. There is evidence which indicates that it is the aluminium oxide that provides the hydrophilic surface for aluminium, hence mechanical or chemical damage to the oxide layer should be avoided. It is also known that the chemical treatment of metals can change the affinity of the surface. For example, copper which is normally hydrophobic can be made hydrophilic with a particular chemical treatment.

Aluminium reacts readily with hydrochloric acid giving hydrogen, more slowly with sulphuric and only very slightly with nitric acid. The etching solution used in making 'deep etch' aluminium litho plates normally includes ferric chloride, cupric chloride, and hydrochloric acid.

MAGNESIUM

Magnesium has been called the lightweight champion of metals since it combines a low specific gravity (1·74) with a relatively high strength. This strength may be doubled or even trebled when small amounts of other metals are blended with magnesium. The development of the jet aircraft was largely made possible by these magnesium alloys. In 1943 the world production of magnesium was 240000 tons compared with 1000 in 1920 and 20000 tons in 1937. At the end of the war in 1945 new uses for magnesium were sought in an attempt to absorb the enormous quantities of magnesium which could be produced. The printing industry provided one of these new applications in

23

letterpress line blocks, produced by the new process of powderless etching. Subsequently this technique, initially developed for magnesium, was to be very successfully applied to the production of zinc and copper blocks.

Magnesium is rather a brittle metal. It can be etched more rapidly than zinc because it is more chemically reactive, but for the same reason it is more subject to corrosion. Corrosion need not be a serious drawback to the use of either metal providing that precautions are taken in the storage and use of the blocks. Magnesium is attacked by dilute acids and by several salt solutions but unlike aluminium and zinc it is quite stable to alkalis.

Magnesium is commonly associated with the brilliant white flame of burning magnesium ribbon or with its contribution to certain fireworks. It may seem odd that the same metal could be used to make printing plates and frying pans. The explanation is that magnesium will burn only when melted in contact with large supplies of air, so that the finely divided forms of the metal provide the greatest fire risk. Magnesium is rapidly etched by nitric acid.

CHROMIUM AND NICKEL

The outstanding properties of these two metals are their corrosion resistance, their attractive appearance and their hardness (Table 2.3). Their main uses are either in alloys or as thin electrodeposited coatings on the surface of other metals. As an example of an alloy containing the two metals, stainless steel with 18% chromium and 8% nickel has outstanding corrosion resistance combined with strength, toughness and relative ease of fabrication. Mention has already been made of the use of stainless steel as a component of bimetallic litho plates. As thin coatings on other metals it is common practice to lay chromium on an intermediate layer of nickel. In order to improve the wear and corrosion resistance of printing surfaces coatings of one or both of the two metals are electroplated on to stereos, electros, bimetallic litho plates and gravure plates and cylinders.

3. Printing alloys

Most printing surfaces are still made of metal even in this 'plastic age'. These surfaces are either prepared by some treatment of the metal plate or cylinder, normally including etching, or by casting molten metal into the required shape. For the casting process, a metal is needed which will melt at a fairly low temperature, flow easily into the mould through small pipes and nozzles to give an accurate reproduction of type and illustration, and also wear well on the printing machine. Unfortunately no single metal meets these conditions satisfactorily so it is necessary to use a combination of metals known as an alloy. Printing alloys are a combination of three metals, lead, tin and antimony.

Table 3.1 Specific gravities and melting points

	Lead	Antimony	Tin
Specific gravity	11·37	6·71	7·29
Melting point °C	327	630	232

Lead melts at a low temperature and is very easily cast, but it is far too weak to be used alone as a printing surface. Although one of the densest of metals, it is extremely soft and malleable. Lead sheets can be produced by a light rolling and lead pipes or tubular containers can be extruded through a die. When freshly cut, lead has a bright metallic surface, but in contact with the air this soon returns to the dull bluish grey colour, so characteristic of the metal. Even so, it has good resistance to corrosion, a point that was obviously appreciated by those early plumbers, the Romans, for lead pipes have been excavated at Pompeii, Rome and Bath and found to be in good condition. Lead forms many useful alloys, apart from those used in printing. These include the range of solders, containing varying amounts of tin, and alloys with antimony which are used in accumulators.

Antimony is a bluish white metal with a brilliant lustre. Its highly crystalline structure makes it extremely brittle, and when struck with a hammer it will shatter into small fragments. For this reason, the metal is rarely used alone in industry but it finds many uses in alloys, most of which also contain lead. Advantage was taken of the brittleness of antimony when, particularly in the first world war, it was alloyed with lead to give a material which shattered into fragments of shrapnel on impact. Normally antimony is alloyed with lead in order to improve hardness and strength in sheets, pipes and accumulator plates. It is sometimes stated that the high quality of type castings is due to the fact that antimony expands on cooling. In fact, this expansion is insufficient to overcome the considerable contraction of lead, but a high antimony content in the alloy does help to improve the accuracy of casting.

Tin is soft, malleable and ductile but much tougher than lead. It has an attractive appearance, excellent resistance to corrosion and can be safely used in contact with most foods. This combination of properties makes tin a very useful metal in packaging, particularly when coated on to sheet steel. The thickness of the tin coating on tin plate is of the order of 1–2μm and only about 1·4% of the weight of a 'tin can' consists of tin. In printing alloys, tin contributes toughness and wear resistance in combination with the antimony. Other alloys containing tin include bearing metals, solders, pewter, bronze and phosphor bronze.

ALLOYS

Alloys are distinctive materials, each one having its own individual character and properties. They are not simply mixtures of two or more metals, having the average properties of the metals present. For example, 'Duralumin', probably the best known of all aluminium alloys, contains 3·5–4·5% copper, 0·5% each of magnesium and manganese, together with small quantities of iron and silicon, yet its mechanical qualities are far superior to those of any of its individual constituents. Again, tin is a very soft metal, yet in type alloys, if the percentage of tin is increased the hardness of the alloy may be improved. Clearly the properties of an alloy need not be related to the properties of the various constituent materials.

An understanding of the nature of a piece of cast type must be based on a knowledge of the changes which take place in the alloy as it is cooled from its molten condition to become a solid.

When a pure metal in its molten state is cooled, it solidifies at a fixed sharp temperature. If, for example, molten lead is allowed to cool, its temperature will fall steadily to 327°C and be held there for a period while the metal solidi-

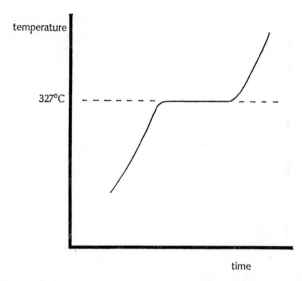

3.1 Cooling curve for lead.

fies. When solidification is complete the temperature will then continue to fall at a steady rate. This can be shown very clearly by drawing a cooling curve, relating the temperature of the lead to the time of cooling (fig. 3.1). In a similar way, if solid lead is heated its temperature will steadily rise to 327°C, where it will remain steady until melting is complete and the temperature rises again. The fixed temperature of transition between the solid and the liquid state may be called the *melting point* or *freezing point* of the metal.

These terms, melting point and freezing point, cannot be applied to alloys because melting and solidification usually take place over a range of temperature. On cooling the molten alloy there is one temperature at which the alloy starts to solidify called the *liquidus temperature* and a lower temperature at which the alloy becomes completely solid called the *solidus temperature*. A cooling curve for an alloy shows these two temperatures and the *melting range* that lies between them (fig. 3.2).

Alloys may be considered as solutions of different metals in each other. For example, in type alloys, tin and antimony are dissolved in lead, and in duralumin, copper and other materials are dissolved in aluminium. Because these metals are dissolved, not simply melted, an alloy may contain metals which are in a fluid condition at a temperature much below their melting point. For example, although antimony melts at 630°C, most printing alloys are completely molten above 300°C. Perhaps this effect is better understood if we remember

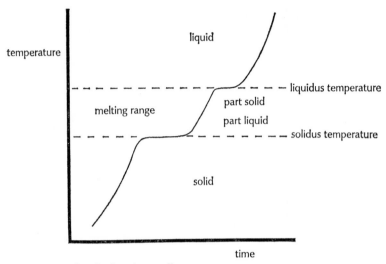

3.2 Cooling curve for a lead–antimony alloy.

that although common salt has a melting point of 801°C, it will readily dissolve in water at ordinary temperatures.

When salt is dissolved in water, the resulting solution has a lower freezing point than the original water, and use is made of this fact when icy roads are salted in winter time. Just as dissolved salt lowers the freezing point of water, so the dissolving of one metal in another lowers its freezing point or, more correctly, its final solidus temperature. If one continues to add salt to water, a point is reached when the freezing point is longer lowered. At this point, the mixture contains 23·6% of salt and is known as the *eutectic mixture*. Its freezing point of − 21°C is lower than that of any other mixture of salt and water.

In a similar way, for any alloy system, there is a certain *eutectic mixture* of the metals present which has a lower solidus temperature than any other mixture of these metals. For example, the eutectic mixture for alloys of lead and antimony contains 88% lead and 12% antimony and its solidus or eutectic temperature is 252°C. When any molten alloy of lead and antimony is cooled, final solidification always takes place at this eutectic temperature of 252°C. For any alloy, other than a eutectic mixture, solidification starts at a higher temperature and continues over a 'pasty range' until the eutectic temperature is reached (fig. 3.3). Only in the special case of the eutectic mixture does the alloy have a single freezing or melting point like a pure metal.

The eutectic mixture for printing alloys consists of 4% tin, 12% antimony and 84% lead, and the eutectic temperature is 239°C. This is a low melting point when compared with the average of the melting points of the constituent metals, lead (327°C), antimony (630°C) and tin (232°C). Examination under the microscope shows that this eutectic alloy has no single crystalline form but consists of a laminated structure of the three metals laid side by side. Unfortunately it is rather soft for use as a printing surface, and so where possible printing alloys contain greater proportions of tin and antimony in order that some harder crystals solidify during the pasty stage.

If a typical printing alloy, for example, one designed for 'Monotype' composition containing 7% tin, 15% antimony and 78% lead, is cooled from a temperature of 350°C, three quite distinct stages can be identified. In the first stage the rate of cooling remains steady for as long as the whole of the mixture is liquid. At about 260°C the *separation point* is reached, when the rate of cooling slows down and very small solid particles begin to form in the molten metal. These particles become larger and more numerous as the alloy thickens to a pasty consistency. The actual nature of the particles depends on the proportion of each metal present. In the case of this particular alloy they are tin-

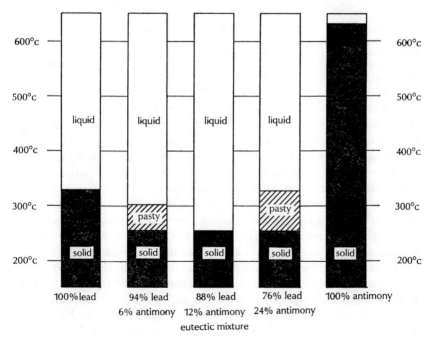

3.3 Solidification temperature of some lead–antimony alloys.

antimony crystals, cubic in shape, very hard and much lighter than the surrounding alloy. When the temperature reaches 239°C the proportions of tin and antimony have fallen to the eutectic proportions of 4% tin, 12% antimony and 84% lead, and the temperature remains constant while this eutectic mixture solidifies around the tin-antimony crystals. In the third and final stage of cooling the temperature of the solid alloy falls steadily to room temperature. The hardness of a printing alloy depends on the amount of tin-antimony crystals formed during the second stage of cooling, so it is desirable to have a high proportion of these present. On the other hand, the separation point rises with the amount of antimony present and thus the harder alloys have to be worked at a higher temperature. It is common practice, therefore, to use softer alloys with a near-eutectic composition in slug casting since the machine conditions demand an alloy which will solidify quickly. Single letter casting machines can be run at much higher temperatures and speeds of casting, and so harder alloys can be used. The compositions of some typical printing alloys are shown in Table 3.2.

Table 3.2 Examples of printing alloys

| Application | Composition | | | Melting range |
	Tin	Antimony	Lead	°C
Slug casting	4	11	85	239–247
Stereotyping	6	15	78	239–261
'Monotype' casting	10	16	74	239–274

In each case the working temperature of the metal must be kept well above the separation point, because it has to remain completely molten for the whole journey from the melting pot to the face of the mould in which it is to be cast. If separation starts on the journey then the nozzles of the pouring apparatus will become choked and the machine will cease to function properly, or at the least, a poorly defined face will appear in the casting.

Where alloys contain more than 4% tin and 12% antimony the first crystals to separate out are composed of equal proportions of these two metals. If the alloy is held for a long time in this semi-liquid condition, these tin-antimony crystals will float to the surface, because they are much lighter than the molten metal around them. A tin content of 4% or less in an alloy with a high proportion of antimony reduces the tin content of the separating crystals to about 10%. This type of crystal is long and narrow, but it behaves very similarly to the cubic type mentioned earlier. When the tin and antimony content of an alloy falls below the eutectic composition, the first metal to solidify out will be lead, and this will tend to sink to the bottom of the mixture because its specific gravity is higher than that of the surrounding liquid. All printing alloys except the softest ones used on slug casting machines are essentially made up of a large number of tin-antimony crystals held together by the eutectic alloy around them acting as a cement. A large number of small crystals evenly distributed through the eutectic will give a more homogeneous and better wearing material than a small number of large crystals with large areas of the soft eutectic exposed in between. The size of the crystals formed depends mainly on the rate of cooling. Slow cooling allows the crystals time to grow large and to concentrate in the upper part of the casting. Fast cooling gives small crystals closely packed together and evenly distributed through the casting.

The structure of a casting can best be examined with a microscope giving × 100 magnification and illumination which falls vertically on the specimen. The casting is cut with a metal saw and the exposed surface polished with successively finer grades of emery cloth, then powdered alumina and finally with a soft cloth. To increase the contrast between the white tin-antimony crystals and the darker eutectic, the polished face is immersed in a 5% solution of silver nitrate, the excess solution being washed off with water.

OXIDATION LOSSES

Printing alloys are melted down after use and used again. During remelting some of the metal is oxidised and floats to the surface of the alloy as a fine powder called dross. If the remelting is carried out at the right temperature for the particular alloy concerned the loss is only from 0·05% to 3% of the total weight of metal. In most of the alloys, however, the proportion of tin oxidised is higher than the average for the metal as a whole, while the amount of antimony oxidised is below the average. Constant remelting would lead to a serious change in composition of the alloy, particularly a loss of tin. It is customary, therefore, to add a small amount of metal with more than the normal proportion of tin at each remelting to maintain the tin content at its proper level. In addition, the metal is *assayed*, that is, its composition is measured, from time to time, so that exactly the right amount of each metal may be added to bring their proportions back to the right figure. Most metal suppliers offer a free assay service to their customers.

IMPURITIES

The most common impurities which find their way into printing alloys during remelting are zinc and copper, dissolved into the alloy during remelting from scraps of brass rule, zinc, or copper plates. The ease with which zinc oxidises causes the formation of a skin of oxide on the surface of the molten alloy. This skin, which is very strong, will lead to heavy losses of metal in the dross. Zinc also reduces the fluidity of the molten alloy and prevents it from flowing properly into the tiny crevices of the mould during casting. The presence of zinc in an alloy shows itself by very rapid dulling of the surface when the top of a batch of molten metal is skimmed. Little more than 0·001% of zinc is needed to affect the working qualities of a printing alloy.

In very small quantities, copper does not seriously affect printing alloys and it is sometimes added deliberately to typefounders' alloys to increase their hardness. But at the comparatively low temperatures used in sorts and slug casting, crystals of copper and antimony combined freeze out before the rest of the alloy begins to solidify. Such crystals choke the outlets from the metal pots and impede the flow of metal to the mould. Nickel is a less common impurity with a similar effect on the alloy to that of copper. Its effects become obvious, however, when about a hundredth of one per cent is present.

A SIMPLE EQUILIBRIUM DIAGRAM

We have already seen how the liquidus and solidus temperatures for a particular alloy can be found by plotting a cooling curve of temperature against time. If these temperatures are obtained for a series of alloys containing two metals in various proportions, an *equilibrium or constitution diagram* may be produced. An equilibrium diagram for lead-antimony alloys is shown in fig. 3.4. The far left of the graph represents 100% lead, the far right 100% an-

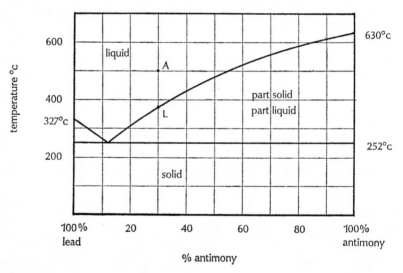

3.4 Equilibrium diagram for lead–antimony alloys.

timony and the points in between, alloys of the two metals in all possible proportions. The vertical axis represents temperature so that any point on the diagram represents a particular lead-antimony alloy at a particular temperature, *eg* the point A represents an alloy containing 70% lead and 30% antimony at a temperature of 500°C. If this alloy is cooled, at 380°C (point L) it will reach its liquidus temperature when antimony crystals form and begin to grow in size. By the time a temperature of 252°C has been reached, the proportion of antimony in the molten alloy has fallen to 12%, *ie* eutectic proportions, so that at this temperature all the remaining liquid solidifies around the antimony crystals. The constitution diagram illustrates the fact referred to earlier, that the only alloy with a single melting or solidification point is one containing eutectic proportions of the constituent metals.

THE STRUCTURE OF SOLID ALLOYS

This equilibrium diagram for lead–antimony alloys is extremely simple because in cooling from the molten state, only two materials solidify, firstly antimony crystals and secondly a eutectic mixture of lead and antimony. The situation may be much more complex with other alloys, because several different compounds of two or more metals may solidify at different cooling stages. One such intermetallic compound which plays an important part in type alloys contains 50% antimony and 50% tin. Another possibility is that the constituent metals remain in solution in the solid state. The composition of these so-called *solid solutions* may vary within certain limits.

To sum up, when a molten alloy is cooled the substances which solidify may include:
(a) One of the metal constituents
(b) An intermetallic compound of the two metals, having a fixed composition and melting point
(c) Constituent metals in solid solution, the composition of which may vary within limits
(d) A eutectic mixture of two or more ingredients with a fixed composition and melting point.

TIN–ANTIMONY ALLOYS

The equilibrium diagram for tin–antimony alloys (fig. 3.5) is more complex because the system involves the solidification of solid solutions as well as intermetallic compounds. Where several different substances solidify from an alloy it is customary to label them with Greek letters. For example, in this particular system, the α constituent is a solid solution containing up to 11% of tin, and the β constituent is substantially the intermetallic compound

33

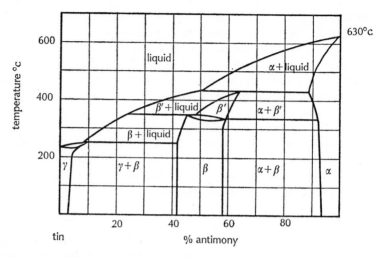

3.5 Equilibrium diagram for tin–antimony alloys.

containing 50% tin and 50% antimony with small amounts of tin or antimony in solid solution.

Alloys containing between 58% and 90% of antimony first deposit crystals of α on cooling but when the temperature reaches 425°C a reaction occurs which results in some of the α changing to the β-constituent. The solid alloys consist of a mixture of the two constituents.

The pure β constituent is formed from alloys containing 42% to 58% of tin. The intermediate compound β' which is shown in the diagram is purely of theoretical importance; on cooling it changes to β as the result of a rearrangement of the atomic structure.

THE TERNARY SYSTEM, TIN–ANTIMONY–LEAD

So far we have considered equilibrium diagrams for binary alloy systems containing only two metals. Matters become more complicated when the system contains three metals. It is not possible to depict completely in one diagram the effects of changes in both composition and temperature.

In order to express the composition of all alloys containing the three metals, it becomes necessary to use a triangular diagram (fig. 3.6). Each corner of the triangle represents one of the pure metals. Each side of the triangle represents the compositions of the alloys containing two only of the metals. Each point within the triangle represents an alloy containing all three metals. The diagram is divided into areas according to the nature of the

34

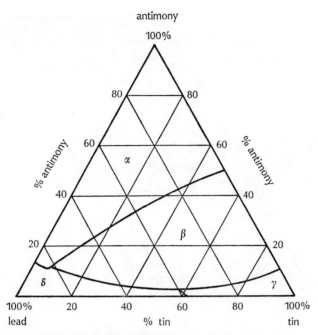

3.6 Equilibrium diagram for tin–antimony–lead alloys.

constituents which are first to crystallise when the alloys cool from the liquid state. The constituents which are present in printing alloys are

(a) solid solution consisting substantially of lead;
(b) antimony-rich solid solution;
(c) solid solution based on the tin-antimony compound, 50% tin, 50% antimony.

The lead-rich corner of this diagram, which covers printing metals, is shown enlarged in fig. 3.7. Here a series of 'contour' lines enable the liquidus temperature of any alloy to be found.

If the contours are followed from B representing pure lead to X, it will be seen that the liquidus temperature falls. In other words, the addition of antimony and tin to lead reduces the liquidus temperature. The alloys in this range deposit crystals of lead (δ) in the initial stages of solidification.

X is the ternary eutectic point, having the lowest liquidus temperature of the alloys within the field.

From X to the line AD, the liquidus temperatures rise again, steeply, as shown by the closeness of the contours. These are antimony-rich alloys, depositing α solid solution on solidifying.

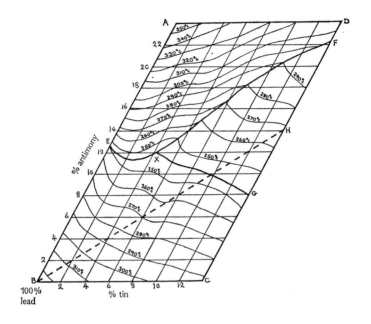

3.7 Diagram of first freezing points for tin–antimony–lead alloys (Weaver).

In the field X G F, the rise in liquidus temperature from X towards F H is much more gentle. The diagram thus shows that the melting point is increased much less when tin and antimony are added together than when antimony is added alone. Alloys in this field first deposit the hard tin-antimony crystals when they solidify.

The ternary diagram thus gives a good deal of information about the primary crystallisation of the alloys. It does not indicate the last constituent to solidify – but in printing metals this is always the same, the ternary eutectic. It has been explained that, in lead-antimony alloys free from tin, the eutectic is an intimate mixture of fine crystals of lead and antimony. The ternary eutectic is similar in nature but more complex. It has the approximate composition of X – 4% tin, 12% antimony, 84% lead, and is a finely divided mixture of the three phases enumerated above – α, β and δ.

The ternary eutectic is the final constituent to solidify in virtually all the alloys above the line B H in the diagram.

4. Polymers

Polymers are so much a part of printing that it would be impossible to discuss the composition and properties of printing materials without first considering the nature of these giant molecules. Some polymers are printing materials in their own right, for example, as plastics they are printing plates, films for page planning and plate mounting, and packaging materials which may have to be printed. Many more polymers are components in printing materials, for example, paper largely consists of the natural polymer cellulose, photographic emulsions are normally coated on to a polymer film base, the light sensitive coatings and photo-resists used at the printing-down stage of platemaking are all based on some form of polymer, and natural or synthetic resins are the film-forming ingredients in many printing inks and surface coatings.

The word 'poly-mer' means 'many parts' and polymer molecules are built up when large numbers of small chemical units are linked together in long chains or networks. If we represent the repeating unit by the letter O, part of a long chain polymer can be represented as

$$-O-O-O-O-O-O-O-O-$$

The unit O is called the monomer and normally consists of a small number of atoms, eg the gas ethylene C_2H_4, with two atoms of carbon and four atoms of hydrogen.

The reaction in which large numbers of monomer molecules link together is called polymerisation and the end product of the reaction a polymer. The monomer ethylene, when subjected to high temperatures and pressures, links up or polymerises to form the familiar plastic material polyethylene or polythene, a molecule made up of large numbers of the repeating unit $-C_2H_4-$

$$- C_2H_4 - C_2H_4 - C_2H_4 - C_2H_4 -$$

Since monomers are small molecules consisting of few atoms they are generally gases like ethylene or liquids like styrene. On the other hand, polymers are giant molecules built up of between 200 and 2000 monomer units. They

37

are, of course, solids whose valuable mechanical properties are due to their large molecular size. Whether the polymer is rubber-like, plastic or fibrous depends on the actual shape of the molecule. The word *plastic* is loosely used to cover almost any manufactured form of a polymer, *eg* film, tubing, moulded articles, etc. More correctly a plastic is a solid material with appreciable mechanical strength which at some stage in its manufacture can be cast, moulded or polymerised directly to a shape. When polymers are produced in granule or powder form for use in surface coatings they are generally termed *resins*, although precisely the same chemical material produced, say, as film would be described as a plastic. The group of polymers known as *rubbers* are those materials capable of being stretched to at least double their length and then able to return rapidly to substantially their original length.

THERMOSETS AND THERMOPLASTICS

Plastics can be divided into two main groups, thermosets and thermoplastics. Thermoplastics are materials which have the property of softening repeatedly on the application of heat and of hardening again on cooling. Thermosets soften only once on the application of heat, then harden irreversibly. This process in which thermosets polymerise and harden is called 'curing'. Thermoset plastics are insoluble in solvents, whilst thermoplastic materials are generally soluble, or at least swell in certain solvents.

This difference in properties between thermosets and thermoplastics is explained by considering their molecular structure. If we again represent the repeating unit of the polymer by the symbol O, then molecules of a thermoplastic can be represented by long chains.

$$-O-O-O-O-O-O-$$
$$-O-O-O-O-O-$$
$$-O-O-O-O-O-O-$$

Since there are no bridging links between these long chains, when thermoplastics are heated the neighbouring chains can slide over one another and the plastic will soften and be able to flow. Solvent molecules will also be able to find their way in between the chains, and may move them apart to make a perfect mixture of polymer molecules and solvent molecules, *ie* a solution of the polymer in the solvent.

Thermosetting plastics polymerise to form large networked molecules. In practice the network is much less regular and linked in three dimensions although the diagram can only show it in two. Such an arrangement is obviously far more rigid than that of separate chains and so we find that after polymerisation, this type of material is unable to flow under the action of

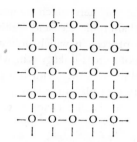

heat. Solvent molecules can make little impression on such a rigid structure and while solvents may sometimes be able to swell a thermoset plastic, they are not able to dissolve it. Polymers used in plastic laminates like 'Formica' are thermosets which illustrate well this excellent resistance to heat and solvents.

The structure of rubber is somewhere between that of a thermoset and a thermoplastic, and it can be considered as a loosely linked network. In the process of vulcanising, rubber is treated with sulphur (S) which forms bridging links between the chains of molecules, thus making the rubber more thermosetting in character and less soluble in solvents (fig. 4.1). By varying the conditions in the vulcanising process, rubbers are produced ranging from the soft to the very hard materials of the type used in battery boxes.

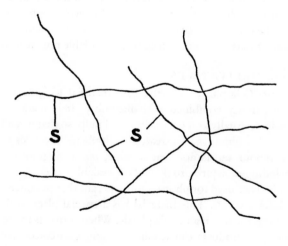

4.1 Sulphur cross linkages formed in the vulcanising of rubber.

As one would expect, the polymers which are used widely as fibres, *eg* Nylon, Terylene, have long chain molecules and are thermoplastic in character.

Photopolymerisation, a process in which polymerisation is triggered by the action of light and photo cross-linkages, in which light causes polymer molecules to further combine together with bridging links, are both used extensively in the production of plates for letterpress and litho. These plate processes are explained in more detail in chapter 15.

NATURAL POLYMERS

One tends to think of polymers as materials of the twentieth century but a very important group of natural polymers were in use long before the plastics industry was born. The group includes rubber, shellac, gum arabic, gelatin, starch, casein, rosin and of course cellulose, without which it is hard to imagine a printing industry.

Cellulose, the chief structural material of the vegetable world, occurs in a pure form in cotton, but it is mixed with many other substances in wood. The various processes of extracting wood pulp for paper manufacture are designed to remove some of the impurities present with the cellulose. These processes are described on page 61. Molecules of the natural polymer cellulose are each built up from about 2000 units of glucose $C_6H_{10}O_5$. In a similar way molecules of natural rubber contain about 5000 units of isoprene C_5H_8.

Cellulose itself has excellent tensile strength but it lacks the solubility needed for use in surface coatings and the flow properties of a 'plastic' material. The versatility of the hydroxyl groups on the cellulose chain molecules makes it possible to introduce various chemical groups and modify the properties of the cellulose to overcome these drawbacks, whilst still retaining the advantage of starting from a cheap freely available raw material.

CELLULOSE DERIVATIVES

Cellulose film or 'Cellophane' was the first material to give the packaging industry transparency combined with flexibility. It is made by chemically treating cellulose, usually as it occurs in wood pulp, so that it can be dissolved and regenerated in film form. Untreated 'Cellophane' provides a poor barrier to moisture vapour and is not heat sealable, but a range of coated films is available including moisture-proof and heat-sealable varieties.

The first film base used for photography was another polymer derived from cellulose, *Cellulose nitrate*. This material has a special place in the history of plastics for it was the first man-made plastic. When cotton is treated with nitric acid a number of products can result including gun cotton (an explosive), collodion (the solution formerly used in the wet-plate process of photography), and the polymer cellulose nitrate or nitro cellulose, as it is more often called when used in lacquers or other surface coatings. Although nitro cellulose is still widely used in hard glossy lacquers, and in gravure and flexographic inks, it has largely been replaced as a film owing to its high inflammability and the fact that it tends to discolour and become brittle with age.

The material that replaced cellulose nitrate as a photographic base film was *Cellulose acetate*. This was commercially developed during the early years of the twentieth century. The first world war had created a demand for fireproof dopes for aeroplane wings and after the war new outlets were sought for the large quantities of cellulose acetate being produced. In the years that followed acetate rayon was developed as a textile fibre and cellulose acetate became established as a plastic, used in film, sheet and moulded forms. Cellulose acetate is produced by the action of acetic acid on cotton or wood pulp. As with nitro cellulose, several grades are available, depending on the way the reaction is carried out. One form, cellulose triacetate, has very good dimensional stability with changes in moisture content, an essential property for photographic film, particularly when used in the graphic arts. Apart from being stable and virtually non-inflammable, cellulose acetate has the advantage of crystal clear transparency. This last property has been exploited in packaging and display where it is successfully used for window cartons, for rigid boxes holding chocolates, flowers, etc, and as a lamination on the surface of record sleeves, magazine covers, etc, or in combination with paper, foil and other plastics in the manufacture of sachets for shampoos, sweets, etc. Apart from packaging, acetate is used in making a great variety of products including toys, fountain pens, combs, cutlery and recording tape.

Two other cellulose derivatives with printing applications should be mentioned. Like nitro cellulose, *Ethyl cellulose* is not important as a plastic film but it is widely used as a resin in flexographic and gravure inks. *Carboxymethyl cellulose* (CMC or cellulose gum) differs from these other materials in that it is soluble in water. It is used as a size in papermaking (page 66) and as a substitute for gum arabic in 'gumming up' litho plates.

SYNTHETIC POLYMERS

The materials that we have considered so far are all made from a natural polymer, cellulose, and so they are not truly synthetic polymers. The rapid growth of the plastics industry in recent years has been largely based on the completely synthetic polymers like polythene, polystyrene and nylon. These synthetic polymers are formed by two basic types of chemical reaction, addition polymerisation or condensation polymerisation.

Addition polymerisation is the process by which a polymer is built up by the repeated additions between one type of monomer molecule and the growing polymer, without any other chemical product being produced. In the majority of cases the monomer is unsaturated and has the structure

$$
\begin{array}{cc}
H & H \\
\diagdown & \diagup \\
C & = C \\
\diagup & \diagdown \\
H & X
\end{array}
$$

The addition process involves the breaking of the double bond between the carbon atoms to form a polymer of the form

$$
\begin{array}{ccccccc}
H & H & \left[\begin{array}{cc} H & H \end{array}\right. & & \left.\begin{array}{cc} H & H \end{array}\right] & H & H \\
| & | & | & | & | & | \\
-\,C\,-\,C\,- & -C\,-\,C\,- & -\,C\,-\,C\,- & \text{etc} \\
| & | & | & | & | & | \\
H & X & H & X & H & X
\end{array}
$$

These addition polymers have a long chain structure and hence show thermoplastic characteristics. They include such important plastics as polythene (where X in the above structure = H, a hydrogen atom), polypropylene (X = CH_3), polyvinyl chloride (X = Cl), polyvinyl acetate (X = $CH_3.COO$, the acetate group) and polystyrene (X = C_6H_5).
(See Table 4.1.)

Polythene

Like many other new commercial materials and processes, polythene was born out of a pure research programme. In 1933, ICI was investigating the effect of very high pressure on chemical reactions. One of these involved the gas ethylene, and by chance the powdery white solid polyethylene was obtained. Six years later, in September 1939, the full scale production of polythene was commenced.

There are two basic types of polythene. The original method of preparation, involving high temperatures and pressures produces a polymer with branching chains, which prevent the close approach of neighbouring molecules and results in a polythene of low density. A second method, discovered by Professor Ziegler in the 1950s, employs a catalyst and much lower temperatures and pressures. This method leads to the formation of linear chains which can pack closely together, resulting in a polythene of higher density.

Table 4.1 Some addition polymers

Name of monomer	Formula	Name of polymer	Repeating unit
Ethylene	$\begin{array}{c} H \quad\quad H \\ \diagdown \quad\quad \diagup \\ C\!=\!C \\ \diagup \quad\quad \diagdown \\ H \quad\quad H \end{array}$	Polyethylene	$\left[\begin{array}{cc} H & H \\ \mid & \mid \\ -C\!-\!C\!- \\ \mid & \mid \\ H & H \end{array}\right]$
Propylene	$\begin{array}{c} H \quad\quad H \\ \diagdown \quad\quad \diagup \\ C\!=\!C \\ \diagup \quad\quad \diagdown \\ H \quad\quad CH_3 \end{array}$	Polypropylene	$\left[\begin{array}{cc} H & H \\ \mid & \mid \\ -C\!-\!C\!- \\ \mid & \mid \\ H & CH_3 \end{array}\right]$
Vinyl chloride	$\begin{array}{c} H \quad\quad H \\ \diagdown \quad\quad \diagup \\ C\!=\!C \\ \diagup \quad\quad \diagdown \\ H \quad\quad Cl \end{array}$	Polyvinyl chloride (PVC)	$\left[\begin{array}{cc} H & H \\ \mid & \mid \\ -C\!-\!C\!- \\ \mid & \mid \\ H & Cl \end{array}\right]$
Vinyl acetate	$\begin{array}{c} H \quad\quad H \\ \diagdown \quad\quad \diagup \\ C\!=\!C \\ \diagup \quad\quad \diagdown \\ H \quad\quad OOCCH_3 \end{array}$	Polyvinyl acetate	$\left[\begin{array}{cc} H & H \\ \mid & \mid \\ -C\!-\!C\!- \\ \mid & \mid \\ H & OOCCH_3 \end{array}\right]$
Styrene	$\begin{array}{c} H \quad\quad H \\ \diagdown \quad\quad \diagup \\ C\!=\!C \\ \diagup \quad\quad \diagdown \\ H \quad\quad C_6H_5 \end{array}$	Polystyrene	$\left[\begin{array}{cc} H & H \\ \mid & \mid \\ -C\!-\!C\!- \\ \mid & \mid \\ H & C_6H_5 \end{array}\right]$

Both these forms of polythene have the advantages of good water and chemical resistance, lightness and excellent electrical properties. They differ in that the low density film has greater toughness and flexibility whilst the high density form has greater rigidity, temperature and grease resistance, and lower permeability to water vapour and gases. Unlike the low density form, high density polythene has a softening point above that of boiling water. This property will allow articles to be steam sterilised, so opening up many applications previously closed to polythene.

Polythene is certainly a versatile material but by far its biggest outlet is film, and most of this is used by the packaging industry. It is widely used in the packing of fresh fruit and vegetables. Film bags prevent the produce drying out but leave it attractively displayed. Apart from food, polythene film is used for packing textiles, hardware, laundry and dry cleaning, chemicals, horticultural products and numerous other materials.

Polypropylene

This newer thermoplastic material discovered in 1954 has a most valuable combination of properties. The basic chemical unit in long chain molecules of polypropylene is not very different from that in polythene (Table 4.1).

Attempts to polymerise propylene were unsuccessful until a method similar to that used by Ziegler in making high density polythene was tried. The result was the lightest plastic so far produced (specific gravity 0·90) and one that was tough, rigid and with excellent heat and chemical resistance.

Polypropylene film is available in stretched (oriented) or unstretched (unoriented) forms. The stretching process may be in either the machine or the transverse direction or in both. When stretched in both directions (biaxially oriented), the tensile properties are roughly the same in any direction and the film is 'balanced'. ICI's Propafilm is a balanced oriented polypropylene (OPP), available in three grades, OS (shrinkable film) and C (coated film).

The process of orientation thins and stiffens the film, and has the remarkable effect of increasing its tensile strength by a factor of approximately four. At the same time it reduces the limpness associated with many plastic films including polythene, and also improves its moisture vapour characteristics. OPP provides a more efficient barrier to moisture vapour and gases than does polythene of the same thickness. Oriented polypropylene film is now directly competing with coated cellulose film and with polythene in many fields of packaging. It is also used as a decorative lamination on printed papers and boards.

Quite apart from its applications in film form, polypropylene is now used in the manufacture of a great variety of articles including tableware, ropes, luggage, toys, safety helmets, drain pipes and printing plates. It is one of the plastics successfully used in the injection moulding of duplicate printing plates.

Polyvinyl plastics

Polythene and polypropylene are both closely related to the very large family of polyvinyl polymers. The various members of the family each have a different atom or group of atoms filling the position X in the repeating unit

$$
\begin{bmatrix}
& \text{H} & \text{H} & \\
& | & | & \\
- & \text{C} - \text{C} & - \\
& | & | & \\
& \text{H} & \text{X} &
\end{bmatrix}
$$

Polyvinyl chloride or PVC, where a chlorine atom fills the position X, is a familiar material with a wide range of applications, from garden hoses to floor tiles, and shampoo packs to corrugated roofing. Apart from its uses in film packaging and plastic binding, PVC is widely used in duplicate plate making. By varying the proportion of plasticiser (a high boiling point solvent which lubricates the polymer chains), it is possible to devise grades of PVC compounds ranging from the completely rigid plate materials suitable only for flat-bed machines, to the flexible forms required in rotary printing plates and in plastic binding. *Polyvinyl acetate* is a particularly important material as an emulsion, where it forms the basis of many white glues used in bookbinding and packaging and where its thermoplastic properties allow its use as a heat sealable adhesive. PVA emulsions are also commonly used as a base for emulsion paints.

Polystyrene is now one of the major thermoplastics and in its various forms it plays an important role in packaging. Polystyrene sheet is heated and vacuum formed on moulds to produce cartons, trays and other containers. Expanded polystyrene is being increasingly used in protective packaging and as an insulating material in building and home decorating.

Polyvinyl alcohol with a hydroxyl (—OH) group in the position X, is an interesting synthetic polymer in that it can be dissolved in water. Sensitised with a dichromate salt, polyvinyl alcohol is now widely used in photoengraving in place of natural polymers like shellac and fish glue.

Condensation polymerisation

Condensation polymerisation is brought about by successive reactions between two chemicals producing a growing polymer and in addition a small molecule such as water or hydrogen chloride. The resulting polymer may have a chain or network structure depending on the number of sites that are available for linkage on the reacting molecules. A simple condensation reaction takes place when, for example, amyl alcohol reacts with acetic acid to form the ester, amyl acetate, the solvent with the familiar smell of peardrops used in balsa cements. This reaction occurs because the OH (hydroxyl) group always present in an alcohol, links with the H (hydrogen) of the acid to form H_2O (water). If we represent the alcohol by the symbol ●—OH and the acid as □—H then the reaction can be pictured as

$$●—OH + □—H \longrightarrow ●—□ + H_2O$$
$$\text{alcohol} \qquad \text{acid} \qquad \text{ester} \quad \text{water}$$

Now, if the alcohol instead of having one OH group has *two* and if the acid also has two H atoms available, the ester link can be made many times so that a very large molecule is formed. We can represent this

$$HO—●—OH + H—□—H \longrightarrow —●—□—●—□—●—□— \text{ etc}$$

| *di*hydric | *di*basic | polyester | $+ H_2O$ |
| alcohol | acid | | water |

In this way it is possible to prepare large *polyester* molecules. When alcohols and acids each with two active groups are used the result will be long chain polymer molecules. As we have seen earlier, these chain polymers have thermoplastic properties, since the application of heat allows the chains to slide over one another and the plastic is able to flow.

Polyethylene terephthalate is an example of a long chain condensation polymer of this type, although it will be more familiar under its trade names as a fibre (ICI Terylene) or as polyester film (ICI Melinex) which has many applications in the printing industry. The reaction of terephthalic acid with ethylene glycol is as follows:

$$HO—CH_2—CH_2—OH + HO.OC—\langle\ \rangle—CO.OH$$
$$\text{ethylene glycol} \qquad\qquad \text{terephthalic acid}$$

$$\downarrow$$

$$\text{etc}—O—CH_2—CH_2—O.OC—\langle\ \rangle—CO.O—\text{etc} + H_2O$$
$$\qquad\qquad\qquad\qquad\qquad\qquad\qquad\qquad\qquad\qquad \text{water}$$
$$\text{polyethylene terephthalate}$$
$$\text{(terylene)}$$

Polyester film

Cellulose nitrate (celluloid) was widely used as a base material for photographic film in the early days of plastics. Since that time less inflammable and more stable plastics have taken its place, in particular, forms of cellulose acetate. Photographic film for colour work in the printing industry must have good dimensional stability when subjected to changing temperatures and humidities. In this and many other respects polyester film (eg Melinex) is close to being an ideal carrier for photographic emulsions. Apart from its stability, the film is colourless, transparent, tough, inert to photographic emulsions and insoluble in most solvents. Polyester film has other roles to play in the production of printing surfaces. It is commonly used as a stable flexible base for film makeup and planning. Compared with glass, it is light, easy to handle and convenient to store.

In the preparation of screen process stencils a photographic positive may be exposed in contact with a photo stencil film which consists of polyester film coated with light sensitised gelatin. Exposure to light hardens the gelatin, and after warm water development the hardened gelatin stencil can be transferred on to the screen.

Polyester film has also found an application in sheet-fed rotary letterpress printing. Flexible printing plates may be mounted with adhesive on to a sheet of polyester film (0·25mm thick) bearing a grid pattern. Since the film is transparent, positioning the plates is relatively straightforward. The polyester sheet is then clamped round the cylinder of the machine.

Other applications of polyester film in the printing industry are in letterpress cylinder dressing, as a drafting film in the preparation of originals from which printing plates or diazo copies are produced, in colour proofing systems and as a base for low cost pre-sensitised litho plates.

Glass reinforced resins

Polyesters can be prepared in a liquid form so that on mixing with a second substance crosslinking rapidly takes place between chain molecules to convert the polymer into a hard tough solid. Resins of this type are frequently reinforced with glass fibre. They have found many successful outlets including boat building, fishing tackle, and as a very useful repair material for the home handyman.

Polyurethane rollers

Polyurethane press rollers are produced by this same process of crosslinking between polyester or similar molecules. A suitable polyester in liquid form and a bridging material (a di-isocyanate) are carefully mixed, led into preheated moulds and cured for several hours. The polyurethane rollers formed are in many ways superior to gelatin composition rollers, being more resistant to mechanical damage and solvents, and more dimensionally stable to changes in relative humidity.

Polyamides

We have seen how polyesters are formed by the multiple linking together of two simple chemical compounds, an acid and an alcohol. Polyamides are formed in a similar way but the compounds this time are an acid and an amine. The most commonly known members of this group of polymers are the *nylons*. Nylons are perhaps best known as textile fibres but they also have important applications in industry, particularly engineering. Nylons may be used to make printing plates, both in photopolymer systems and in duplicate relief plates.

Alkyd resins

If in the condensation polymerisation of a polyester, we choose an alcohol with three — OH groups available for linkage instead of two, then it becomes possible for a polymer network to build up. These networks give the material a more rigid structure and a thermosetting character, *ie* once the polymer is formed, the application of heat will not cause it to flow (page 39).

If, for example, glycerine (glycerol) which has three OH groups, is used in place of ethylene glycol in the reaction with phthalic acid, the result is a thermosetting resin. Polyester resins of this type were given the name *alkyd*, a combination of the words alcohol and acid. The pure alkyd resin produced from glycerol and phthalic anhydride has little value in surface coatings because it lacks solubility in common solvents but when these alkyds are modified with certain vegetable oils, *eg* linseed, soya bean, they form a very important group of vehicles for letterpress or litho printing inks (page 168).

THERMOSETTING PLASTICS

So far we have mainly been dealing with long chain polymers which are always thermoplastic in their behaviour, *ie* whenever heat is applied the separate chains are free to slide over one another so that the plastic is able to flow, take up a new shape and on cooling become hard again. However in the case of the polyesters we saw that if one of the two simple substances linking together to form a polymer had *three* available sites for linkages, then a three dimensional network or lattice could be formed. An arrangement of this type, clearly much more rigid than that of separate chains, is normal for all thermoset plastics. As one might expect, when polymerisation is complete these materials are not softened by further heat or much affected by solvents.

The thermosetting materials used in the production of matrices and dupli-

4.2 Reaction of phenol with formaldehyde.

cate plates are impregnated into a cellulosic or asbestos board. The simple chemical substances from which the polymer forms are usually phenol and formaldehyde; the nature of the reaction is shown in fig. 4.2. In this case the molecules of phenol have three positions available for linkage and the smaller formaldehyde molecules have two.

The polymerisation proceeds in stages, the first stage being the linkage of two molecules of phenol with one of formaldehyde. It is possible to stop the polymerisation at an intermediate stage, and partially polymerised resins of this type, together with additives are incorporated into the phenolic boards used in the production of matrices and plates. The heat and pressure of the moulding press causes further chemical linkage and the formation of large polymer networks which make the boards hard, rigid, solvent resistant and reasonably stable to heat.

To a much lesser extent, phenolic boards are used in the manufacture of printing plates. Whilst they are capable of excellent results, their use is limited by their rigidity, which makes them unsuitable for rotary printing.

Thermoset polymers have found a number of other applications in printing surfaces. Copper clad laminates in which a layer of copper is laminated to a thermoset plastic of the Bakelite (phenol formaldehyde) type have been used to produce fine screen halftone blocks. These are claimed to show less wear than those produced from rolled copper. These copper clad laminates are also used in the manufacture of printed circuits.

Thermoset phenolic plastic is a familiar material for composing room furniture, for example Stephenson Blake's 'Resalite'. This material is claimed to have wearing properties at least equal to those of lightweight alloy and is impervious to climatic change.

COLD CURING

Although the term thermoset implies the use of heat in the build up of large networks of molecules, the same type of polymerisation can be brought about at room temperature, when the process is called cold curing. Epoxy resins are frequently used in this way. The addition of a curing agent or cross-linking agent causes the epoxy resin to polymerise further to become a hard, tough, infusible solid. In practice, the epoxy resin is used in a two part mixture which is made immediately before use. The flexibility of the final product can be varied by including other polymers in the mixture. Cold cured epoxy resins of this type are used in the plastic backed 'college electrotypes'. The shell, which can be considerably thinner than a conventional electrotype, is filled with a mixture of two pastes, the proportions of which can be varied to give differing degrees of flexibility.

PRINTING ON PLASTIC FILMS

The problems faced when printing on plastic films are quite different from those which arise when printing paper. All plastic films offer a smooth continuous surface which does not allow inks to penetrate as they do into paper. However, since the films are all thermoplastic, each of them is likely to be affected by a specific solvent or solvents. The ink maker may be able to include such a solvent to enable the ink to key on to the smooth surface of the plastic.

Cellulose film (eg Cellophane) and cellulose acetate do not present any serious problems when printed by gravure or flexography. However, good adhesion is much more difficult to obtain when printing on the less polar and therefore more inert plastic films, polythene and polypropylene. The problem has been largely overcome by physically or chemically treating the film before it is printed. The most commonly used method is to pass the film through an electronic discharge which, impingeing on the surface of the film, renders it receptive to inks and other surface coatings.

5. Paper and board – raw materials

It would be difficult to exaggerate the importance of the role that paper plays in our modern world. For many years, it has been the chief medium for the communication of knowledge and ideas in a permanent form, so essential to the development of commerce, industry, education and government. We have seen its increasing application as a medium for the protection and display of goods in the packaging industry. Perhaps the next period of development for paper will be in the protection and display of human beings in the disposable clothing industry.

Certainly, without paper, it is hard to imagine a printing industry in anything like its present stage of development. Although there is a growing amount of printing being carried out on plastic and metal substrates, paper is likely to be the printer's most important basic material for many years to come. It is clearly essential for anyone involved with printing in any capacity to have a good understanding of the nature of paper and its constituent raw materials, and be reasonably informed on the processes used in its manufacture and subsequent finishing.

THE NATURE OF PAPER

Paper consists essentially of a mat or web of intermeshed cellulose fibres. The mat is formed when a very dilute aqueous suspension of the separated fibres flows on to a very fine wire mesh so that the water drains through, leaving the fibres to settle together into a felted layer. Most papers also contain a number of non-fibrous materials such as china clay and rosin size.

STAGES IN MANUFACTURE

Papermaking can be considered in four distinct stages:

(1) *The manufacture of cellulose pulp* from wood or some other plant material.

(2) *The preparation of stock for the paper machine* This involves the mechanical treatment of the cellulose fibres in the processes of beating or refining so that they will intermesh to give a paper with the desired properties. The treated

fibres and any non-fibrous materials must also be blended into a dilute aqueous suspension ready for the paper machine.

(3) *The papermaking operation* in which water is progressively removed from the dilute suspension first by drainage through the wire mesh and then by suction, pressure and the application of heat to form a continuous web of dry paper.

(4) *After-treatments or processes which may follow papermaking* including calendering, coating, slitting, and cutting.

As one might expect, there are variables at each of these stages of manufacture which influence the characteristics of the paper finally reaching the printer. However, the second stage, in which the cellulose fibres are treated, is generally regarded as the most crucial in determining the properties of a paper. This is well illustrated by the fact that by varying beating conditions it is possible to produce papers as different as greaseproof and blotting from the same cellulose pulp. Differences between papers will be better understood when we have considered their raw materials and the four stages of manufacture in more detail.

FIBROUS MATERIALS

Cellulose fibres can be regarded as the common building bricks of plant architecture, whether it be in a blade of grass or in the trunk of the largest tree. In a few materials, like cotton and linen, the cellulose exists in a pure form but in most plants it is mixed with appreciable quantities of other materials like lignin. To extend the building analogy, the lignin and similar materials in which the fibres are embedded can be regarded as the mortar binding together the cellulose bricks to make the plant stronger and more rigid. From the papermaker's point of view these non-cellulose constituents of plants are undesirable, hence the development of pulping methods designed to free the individual cellulose fibres from these other materials.

Cellulose fibres suitable for paper making can be extracted from nearly all plants, but the yield of cellulose fibres from most plants is too low to make the process economic. Plant materials for paper making must allow the extraction of good yields of fibre of a suitable quality. The choice of these materials will also be influenced by other factors such as their ease of harvesting, the length of their growing period and their accessibility to the pulp mill and the paper markets. The economic advantages of having pulp and paper mills near to the source of raw plant materials have been well demonstrated in Scandinavia and North America.

Although the classification of paper making fibres which follows shows that there are a number of materials which meet the general requirements, in practice, *more than 90% of the world's paper is now made from wood.*

Table 5.1   Classes of papermaking fibres

Class	*Examples*
Seed hairs	Cotton
Bast fibres	Linen, hemp, jute
Grass fibres	Straw, bamboo, bagasse
Leaf fibres	Esparto, sisal
Wood fibres	Coniferous and deciduous woods

The fibres extracted from these plant materials differ in length, breadth and shape but they are similar in general structure. They take the form of hollow translucent tubes, having closed and often tapering ends. The characteristic appearance under the microscope of fibres from different plant sources is very useful in their identification and analysis.

Average fibre lengths range from 1·0 to 1·5mm for esparto, straw and deciduous woods, to 4mm for coniferous woods and 25 to 30mm for cotton: average widths are from 0·01mm to 0·04mm.

Table 5.2   Dimensions of some papermaking fibres

Fibre	*Average length mm*	*Average width mm*
Cotton	10–50	0·025
Coniferous woods	4	0·025
Deciduous woods	1·5	0·030
Esparto	1·5	0·013
Straw	1·5	0·015

The fibres are grouped together along the length of the stem or trunk in various ways, according to their class. For example, in the straws and grasses separate bundles of fibres are scattered at random, whilst in trees they are arranged in roughly concentric rings, each ring being a year's growth.

There are differences, too, in the proportion of non-fibrous material associated with the fibres in the plant. In the single case of cotton the cellulose is present in an extremely pure form, so the process of making a cotton pulp for papermaking is quite simple. However, all the other papermaking fibres have appreciable amounts of non-fibrous material cementing them together in the plant. The whole object of pulp manufacture is to isolate the fibrous matter, and in so doing, cause the minimum amount of mechanical or chemical damage to the fibres.

Some of the more important examples of plant sources of papermaking fibres are given in Table 5.1.

Cotton fibres, the purest form of natural cellulose, are the seed hairs of the cotton plant. The seeds, covered with these fine hairs, are contained in pods or bolls. When the bolls are ripe they burst open and the hairy seeds are picked. Most of the cotton is spun into thread for textile use. Some of the fibres are too short for spinning and these linters may be used in paper making. Cotton rags provide a larger source of cotton for the paper maker. These may be offcuts from the textile mills or simply used old garments. After cleaning and bleaching, both cotton linters and rags produce an excellent pulp for paper making.

Cotton fibres are flat twisted tubes averaging about 28mm in length and 0·025mm in width. If, in the beating process the fibre length is preserved, cotton will yield strong durable papers, of the high quality required for currency, legal documents, drawing papers and high grade stationery. Hand-made papers are often based on cotton or linen or a mixture of the two.

Linen fibres are contained in the ring of bast tissue just below the surface of the stem of the flax plant. These bast fibres only represent about 5% of the weight of the original plant. As with cotton, the paper maker only gets the rejected short fibres, offcuts and used linen rags.

Linen fibres have similar dimensions to those of cotton. One important property of the bast fibres is the ease with which they will split along their length when suitably beaten. This longitudinal splitting or fraying is called *fibrillation*. The hair-like fibrils on the beaten fibres help the fibres to inter-mesh firmly together and so form a strong sheet. The strength of individual linen fibres and their readiness to fibrillate makes for exceptionally strong and durable papers. Unfortunately the high cost of linen pulps limits their use to the manufacture of high quality papers including those for currency, legal documents and airmail papers.

Hemp and jute are other examples of bast fibres used in papermaking, although they are of relatively little importance in Britain. Hemp, obtained from a shrub grown in India, Russia and America, normally comes to the paper mill in the form of waste ropes, twine and cordage. It is similar in dimensions and behaviour to linen, and is used in the manufacture of strong durable papers which include tissues, cigarette papers and wrappings. Jute fibres are extracted from an Indian plant to produce sacking, and it is in this form that they are usually received by the paper maker. Whilst they are much shorter than linen and hemp fibres, they fibrillate well like other bast fibres and so can be used to produce thin strong papers. Jute fibres are also able to contribute bulk in a manner similar to esparto.

Straw fibres are obtained from the stem of the common cereal plants, particularly from wheat. In Britain, during the last war, wood pulp and esparto grass were in short supply and straw was successfully used as a papermaking fibre. Improved methods of extracting cellulose pulp from straw have since been developed, and several countries have seen the advantages of establishing a straw pulp industry, providing a home-grown paper making material. Although they have sometimes been regarded simply as a substitute, straw pulps have established a special place in paper making. Straw fibres are short tubes pointed at both ends. Although their dimensions are of the same order as fibres from esparto grass or deciduous woods (Table 5.2), they may be distinguished under the microscope by the presence of non-fibrous cellulose material in a variety of shapes and sizes (plate 6). Straw pulps must be carefully beaten to retain fibre length. These pulps give papers with a dense hard finish, having low tearing strength and opacity but good bursting strength and a distinctive 'rattle'. Normally straw pulps are blended with wood pulps to produce thin hard writings, *eg* banks, bonds and also some printings.

Other grasses which provide a source of papermaking fibres include *bamboo*, widely used in India, and *bagasse*, the residue of the sugar cane after the extraction of sugar. These fibres resemble those from cereal straw in their general paper making characteristics.

Esparto grass which grows in North Africa and Spain, is strictly a leaf fibre. In the hot dry climate, the leaf curls up into a wiry tube resembling coarse grass. It is pulled by hand, sorted and then baled for transportation to the paper mill, where a cellulose pulp is extracted. Esparto fibres are short narrow tubes sharply pointed at both ends. As with straw, the fibres are always associated with particles of non-fibrous materials which include small pear-shaped hairs from the inner surface of the leaf. These hairs together with the characteristic appearance of the fibres themselves, make esparto relatively easy to identify under the microscope (plate 5).

Esparto fibres do not fibrillate in the beating process, and being relatively short, their value in paper making is to contribute bulk and compressibility rather than great strength. Esparto is generally used in combination with chemical wood pulp in the production of both printing and writing papers.

Wood Pulp in its various forms is by far the most widely used of all the fibrous raw materials for paper. The pulp may be produced from *coniferous* (cone-bearing) trees or from *deciduous* (broad leaf) trees. Conifers, like the spruce and pine grown in Scandinavia and North America are the chief source of wood pulp. The life cycle of the spruce is about eighty years so vast areas of woodland are cultivated in order to meet the steadily increasing demand for paper and board. A large paper making organisation having cutting rights on say 8 000 000 acres might annually cut an area of 100 000 acres, on which seedlings would then grow until they came to be harvested about eighty years later. Demands for wood pulp call for forest management on a very large scale.

Spruce fibres are 3–4mm long. Having thin walls they easily become flattened and under the microscope resemble thin ribbons (plates 1–3). In the beating process the fibres can be fibrillated but not to the same extent as say linen or cotton. Since they are also shorter than these fibres, care must be taken to retain fibre length, on which the strength of the paper depends. Spruce pulps, either mechanically or chemically produced, are widely used in the manufacture of almost every type of paper.

The fibres from Scotch *pine* are similar to those of spruce, but the wood contains far more resinous material and a different method of extracting the cellulose pulp must be used.

Deciduous trees supply much less wood pulp for papermaking than the conifers but they are becoming increasingly important. Varieties used include aspen, birch, poplar, chestnut and beech. Their fibres are relatively short compared with spruce, varying from 1 to 1·5mm. They are able to contribute softness, bulk and opacity to a paper, in much the same way as esparto grass.

THE STRUCTURE AND COMPOSITION OF WOOD

In the process of plant growth, the sap, consisting of water and dissolved salts taken from the soil, is absorbed through the roots and rises by capillary attraction from fibre to fibre through the stem or trunk and branches to the leaves. Carbohydrates are then formed as a result of a complex series of re-actions, called photosynthesis, involving water and the green pigment chlorophyll from the leaves, carbon dioxide from the air and sunlight. The carbohydrates formed by day in the leaves, are then carried to those parts of the plant where growth is taking place.

The annual growth in the trunk or branches of a tree takes place in a thin layer immediately below the bark known as the cambium. This laying down of a layer of new wood each year, leads to the familiar annual growth rings, from which one can estimate the life of a tree (fig. 5.1).

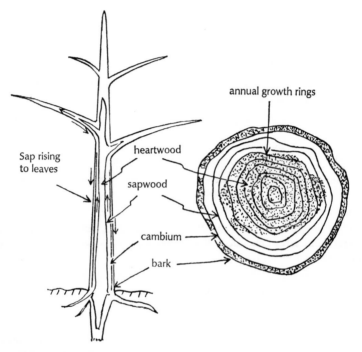

5.1 Gross features of wood.

The cambium increases in size by the repeated division of cells. The layer is not uniform since there is a difference between the cells formed early and late in the growing season. In most woods the cambium consists largely of long thin tubular cells, commonly called fibres.

The structure and composition of these wood fibres has been investigated extensively over many years and modern techniques including electron microscopy have helped to reveal their fine structure. A chemical analysis of wood shows that it contains approximately 50·0% carbon

43·4% oxygen
6·0% hydrogen
0·1% nitrogen
0·5% ash (largely silica)

These elements are combined together in many different compounds which fall into two main groups

5.2 Structure of repeating unit in molecule of cellulose.

(1) *Carbohydrates*, notably cellulose.

(2) *Lignin*, non-fibrous material acting as a cement bonding the cells together.

Cellulose is a white fibrous polymer, containing the elements carbon, hydrogen and oxygen. Its formula can be expressed simply as $(C_6H_{10}O_5)_n$, where n represents the number of linked glucose units. These repeating units have a six membered ring structure (fig. 5.2).

The actual number of units linked together in each molecule depends on the type of plant source and on the severity of the method of extracting the cellulose pulp. Evidence suggests that cellulose molecules in cotton pulp may contain up to 5 000 units, but that the maximum figure for wood pulp is about 2 000. A single cellulose molecule in wood pulp is about 0·75 nm wide and between 800 nm and 5 000 nm long.

These molecules lie side by side along the fibre axis and together form the *microfibrils*, which fray out from the fibres as a result of the beating process. These fine threads are about 10 nm wide and of indefinite length. X-ray studies have also shown that within these microfibrils there are both crystalline and amorphous regions. In the crystalline regions, called *micelles* or *crystallites* the bundles of cellulose molecules are held together in an ordered fashion by links between adjoining chains, whereas in the amorphous regions the molecules are in a disorderly arrangement. Since the micelle regions are much shorter than the cellulose molecules, an unbroken molecule may pass through several micelle regions.

Although the microfibrils can only be studied under an electron microscope, they do associate laterally to form the larger fibrils, which are visible under a high powered optical microscope.

The wood fibre itself appears to exist in four distinct layers surrounding a central cavity called the lumen (fig. 5.3). The outer layer or primary wall (P) consists of a random arrangement of fibrils and contains a high proportion of lignin. In the next three layers S1, S2 and S3, the fibrils are laid in a more ordered way at distinctive angles to the fibre axis. The S2 layer is the thickest and makes up a high proportion of the total fibre weight.

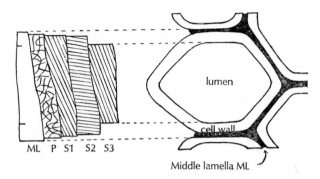

ML   P   S1   S2   S3

lumen

cell wall

Middle lamella ML

5.3 Diagram of a typical spruce fibre cross-section and of its layered fibrillar structure.

THE MANUFACTURE OF CELLULOSE PULP

In all the plant sources discussed so far, the cellulose fibres are bound up with various non-fibrous material. The percentage of cellulose present in the plant varies widely, as is shown in Table 5.3. The object of the pulping process is to

Table 5.3   The cellulose content of some plant sources of papermaking fibres

Plant	% content of cellulose
Cotton	approx 95
Flax (linen)	70
Esparto	46–58
Spruce	55–61
Pine	58
Hardwoods	58–63

separate the fibres from the other unwanted materials, in such a way that the minimum of mechanical or chemical damage is done to the fibres. Since today well over 90% of the world's paper and board is made from wood pulp we will mainly concern ourselves with the methods that have been developed for producing cellulose pulp from wood.

The manufacture of wood pulp

After the trees have been felled and stripped of branches, they have to be carried from the forest to the pulp mill, which is often hundreds of miles away. The logs are taken overland to some mills, but more often they are floated down the fast flowing rivers, swollen by the melting ice and snow in the spring. In the slower waters of the river downstream the logs may be gathered together in huge rafts by a surrounding boom and towed by tugs to a storage point near to the pulp mill.

At the pulp mill the logs are fed into a large revolving drum in which they rub themselves free of bark. The next step in the process depends on the type of pulp to be produced.

Methods of wood pulp manufacture can be considered under the following headings:

(1) Mechanical

(2) Chemical

    (a) Acid (b) Alkaline

(3) Combined mechanical and chemical processes

(1) *Mechanical (or groundwood) process*
In producing mechanical wood pulp no real attempt is made to remove the non-fibrous constituents of the wood. The debarked logs are pressed against a rotating grindstone, the logs being held parallel to the shaft of the stone. Sprays of water cool the grindstone, act as a lubricant and wash away the particles of wood as they are worn away from the logs. The resulting pulp is a mixture of bundles of fibres, broken fibres and in addition all the non-fibrous material that was present in the original wood. The characteristic appearance of a groundwood pulp under the microscope is shown in plate 1.

Newsprint is based largely on this type of pulp, and it illustrates well all its shortcomings. The harshness of the grinding process causes extensive fibre damage and a paper made from the pulp has poor strength. The lignin and other impurities in the pulp result in a paper with a poor colour, which further yellows on exposure to light. To counteract their poor strength and colour, mechanical pulps are normally blended with a chemical pulp. Despite these disadvantages, mechanical pulps have an important role in paper making, firstly because of their cheapness and secondly because newsprint and many other cheap grades of paper and board have a very short working life in which strength, good colour and permanence are not essential properties. Mechanical pulps are much cheaper than those produced by chemical methods, since the yield of pulp from wood is around 95% and no expensive chemicals are used. Although power consumption is high in the grinding process, the pulp mills are normally situated where electrical power is relatively cheap. Mechanical pulp is usually produced from spruce wood.

Increasing amounts of groundwood pulp are now being produced by passing wood chips through a series of refiners (see fig. 6.4 and page 73). The resulting refined groundwood contains a larger proportion of the longer fibres than ordinary stone groundwood, and so it leads to the formation of a stronger sheet of paper. Another advantage of the process is that it is possible to produce groundwood from sawdust or shavings instead of logs.

(2) *Chemical processes*

After the debarking process, logs which are going to be chemically pulped are sliced into small chips about 20mm long and 5mm thick. These chips are then treated with chemical solutions which extract the lignin and other impurities from the wood, leaving the cellulose fibres in a separated and relatively pure state. Several methods have been developed for achieving this objective, some involving acidic solutions and others alkaline solutions.

(a) *Acid process*   In the sulphite process, the only acid process of commercial importance, the wood chips are treated with a solution of calcium bisulphite, containing free sulphur dioxide. The reaction takes place in a digester, a large cylindrical pressure cooker, with an acid-resistant lining. The digester is heated by direct steam at a pressure of about 700kPa, and the process may take up to 20 hours. During this cooking operation, much of the lignin and other non-fibrous materials pass into the acid sulphite liquor. 100 tons of wood chips will yield about 55 tons of cellulose pulp. The remaining black sulphite liquor presents a problem at the pulp mill, since chemicals cannot easily be recovered from it and when discharged to waste it can cause the pollution of a river or lake. Many attempts have been made to find a use for these waste liquors and these have met with some success.

(b) *Alkaline processes*   Like the acid sulphite process, these methods involve cooking the wood chips in a digester, but of course the chemicals used are different and there is no need for the digester to have an acid resistant lining.

In the *Soda process*, sodium hydroxide (caustic soda) is used in the cooking process and pulps are produced from esparto grass, straw and rags as well as from deciduous woods. One important difference from the sulphite method is that chemicals can be recovered from the waste liquor and pulp washings.

In the *Sulphate or Kraft process*, sodium sulphate is added during the recovery stage, but the reactive chemicals in the digester are sodium hydroxide and sodium sulphide. As the sodium hydroxide is used up in the cooking process, more is formed to take its place as a result of the hydrolysis of sodium sulphide. This controlled supply of sodium hydroxide is a great advantage for it avoids drastic alkaline conditions which would attack cellulose and yet allows the digestion to continue at a steady pace. In this way fibre length is retained and a strong pulp results, hence the use of the word Kraft, the German word for strength. Papers made from unbleached Kraft pulp include the familiar brown wrappings and paper bags. The sulphate process is particularly suitable for woods like pine with a large resinous content, which cannot be extracted effectively by the acid sulphite process.

(3) Combined mechanical and chemical processes

In recent years a number of pulping methods have been developed which combine a mild chemical digestion with mechanical disintegration. As one might expect, the product of these processes is intermediate in quality between a mechanical and chemical pulp. In general, the chemicals used are those already mentioned in discussing the purely chemical pulping methods.

The *semi-chemical process* is a two stage process in which a treatment with a normal pulping chemical, *eg* sodium hydroxide or sodium sulphite, opening up the fibrous structure, is followed by mechanical action. In the *mechanochemical process*, plants such as straw or bagasse are agitated at high speed in a hot dilute solution of sodium hydroxide.

Some modern pulping methods, combining chemical and mechanical treatment, have been designed to operate continuously rather than as batch processes. As in many other industries, the trend to continuous operation with the possibilities of automatic control is likely to be the pattern for future development in pulp manufacturing. Ultimately perhaps the wood chips will be processed in stages as they make the journey from the forest to the paper mill by pipeline.

Screening, cleaning and bleaching pulp

After the digested pulp has been washed it still contains impurities that must be removed before papermaking.

Screening is a multi-stage operation in which diluted pulp is made to pass through small holes or slits cut in a series of plates or rotating drums. Useful fibres are able to pass through the holes, whilst larger pieces of material are retained (fig. 5.4). Material removed in this way includes knots, uncooked bundles of fibres and pieces of bark.

The screening process which relies on differences in size is followed by *cleaning*, in which grit and other materials are separated by differences in density. Originally diluted pulp was made to flow over sand traps, into which heavy particles settled and were retained (fig. 5.5a). However, the efficiency of separation has been greatly increased in more recent devices including centrifugal and vortex cleaners. In the vortex cleaners (fig. 5.5b), the diluted stock is forced under pressure into an inverted cone. Good pulp spirals down the wall of the cone and then rises upwards in an inner vortex to leave by a

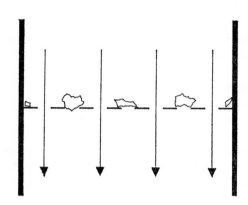

5·4 Screening – separation by differences in size.

central outlet. Heavier particles are carried downwards to the open point of the cone where they are continuously discharged.

As well as improving the colour of a pulp, *bleaching* processes remove more of the remaining impurities and so can be regarded as an extension of digestion. Most industrial bleaching methods are based on chlorine or compounds which supply chlorine, and the commonest of these materials is calcium hypochlorite. At its simplest, the bleaching process can be carried out in a single stage, in which the pulp is treated with a solution of calcium hypochlorite. This method is still used for non-woody materials like esparto and straw. For other substances, bleaching is often carried out in several stages, for in this way it is possible to achieve the most satisfactory balance between the yield, strength and colour of a pulp.

Some lignin remains in a pulp after the earlier digestion process, since the conditions necessary to remove it all would cause fibre damage and hence low yields. The chlorine in the bleaching process has a selective action on the remaining lignin, which renders it soluble in either acid or alkaline solutions. The concentration of available chlorine required depends on the nature of the raw material and on the colour required, and ranges from 2% for an 'easy bleaching' pulp to up to 15% for a Kraft pulp. Bleaching in stages makes it possible to reduce chlorine concentrations at any one stage, and so reduces fibre damage caused by the strongly oxidising action of concentrated bleach solutions in a single stage process. Chlorine dioxide is now widely used in the final stages of bleaching both sulphite and sulphate pulps. Sodium peroxide and hydrogen peroxide have also been introduced for the same purpose, and for the bleaching of mechanical pulps.

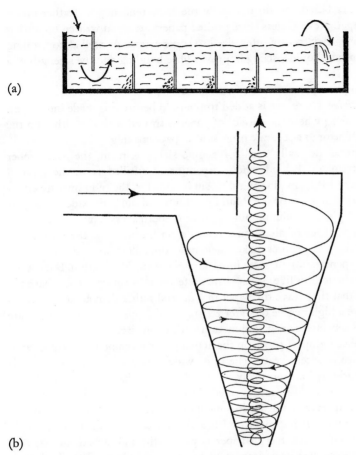

(a)

(b)

5.5 Cleaning – separation by differences in density: (a) sand traps; (b) vortex cleaner.

NON-FIBROUS MATERIALS

Although paper is largely made up of cellulose fibres, it also contains a number of added non-fibrous materials. The most important of these are sizing agents, fillers or loadings and colouring matters.

Sizing agents

Any paper, which is going to be written on, must be sized to prevent the spreading out or feathering associated with blotting paper. Most printing inks are not water based and do not show the same tendency to feather on an unsized paper. Nevertheless since printed papers are often written on during their lifetime, they too are often sized to some extent. As well as controlling the resistance to liquids, sizing improves fibre bonding and the strength of a paper.

The two principal methods of sizing are

(a) *Engine sizing* where size is added to the pulp before it is made into paper.

(b) *Surface sizing* where the surface of paper is treated with size either on the paper machine or as a separate process after papermaking.

The material normally used for engine sizing is rosin, the hard amber coloured resin left behind when turpentine is distilled from the gummy substance exuded from pine trees. Rosin is added to the pulp mixture either as an emulsion in water or as a solution in dilute sodium hydroxide. When this is thoroughly mixed, aluminium sulphate (papermakers' alum) is added to precipitate particles of aluminium rosinate on the walls of the fibres. The coating of size on the fibres makes them more difficult to wet with water, and reduces the penetration of water into the paper by capillary action. For certain papers, synthetic sizes like Aquapel are preferred to rosin size. These have the advantage that they leave the paper in a neutral rather than an acidic condition. Where a high degree of water resistance is required, wax emulsions are used either separately or in conjunction with rosin size.

The surface sizing of paper on the actual paper machine has become common practice in recent years. The most widely used surface sizing agent is starch, a carbohydrate normally obtained from maize or potatoes. Other substances used include carboxymethylcellulose (CMC), polyvinyl alcohol and gelatin. Apart from the improvement in a paper's water resistance, surface sizing also gives an improved finish and reduced fluffing. Tub sizing, in which a manufactured reel of paper is passed through a bath of size as a separate process, was once important but is now only used for high class papers. Surface sizing can also be applied on a calendering unit.

Fillers or loadings

These fine white mineral pigments are important constituents of almost all papers. By filling the spaces between the intermeshing fibres, they improve the printing characteristics of a paper by increasing its smoothness and flatness, its opacity and brightness, and its dimensional stability with changing humidity. Not least, fillers reduce the cost of a paper since they are very much cheaper than fibres. On the debit side, they tend to make papers more abrasive, and so can accelerate the wear of a printing plate. More important, they reduce fibre bonding and the strength of the sheet, so clearly there is a limit to the proportion of fillers that can safely be added. Some imitation art papers may contain up to 30%, but most printing papers normally only carry between 10% and 20% of fillers.

The ideal filler has a very high refractive index and scatters visible light of all wavelengths. It must be chemically inert and available in a finely divided form at a reasonable price. The most widely used material is *china clay* or kaolin, which is obtained from surface deposits in Devon and Cornwall and other parts of the world. *Titanium dioxide* is more expensive but its effectiveness as a filler more than justifies its use. It is produced in two different crystalline forms, Anatase and Rutile (page 157). The high refractive index of both forms (Anatase 2·52, Rutile 2·76) means that titanium dioxide is able to scatter more light and is therefore far more opaque than china clay. Even small proportions of the pigment have an appreciable effect on the opacity of a paper or surface coating. Apart from its high opacity, titanium dioxide has the advantage of being extremely inert chemically. Other fillers which are used in considerable quantities are precipitated chalk (calcium carbonate) and barium sulphate, which in its precipitated form (blanc fixe) is used in producing photographic papers.

Colouring matters

These are added to pulps in the production of coloured papers, but also in smaller quantities when their function is simply to correct the yellowish tint of the pulp and produce a particular shade of 'white'. Clearly, matching the colour of a wet pulp to that of the final dry paper is a difficult task calling for a great deal of skill, coupled with the resources of a laboratory, where preliminary colour matching trials on a smaller scale can be carried out. The colouring matters used in papermaking include those which are soluble in water (dyes) and those which are insoluble (pigments).

In this application, *pigments* may be considered as coloured loadings, filling the interspaces between fibres and having the same influences on the properties of the paper. Both white and coloured loadings have a tendency to produce two-sidedness in a paper because their drainage through the wire leaves them rather more concentrated on the top side of the sheet than on the wire side. Compared to dyes, pigments have better resistance to both light and chemicals. Mineral pigments like ultramarine, prussian blue, the ochres, siennas and chromes are normally used rather than the more expensive organic pigments. Chapter 11 deals with pigments in some detail.

Dyes colour paper in a very different way from pigments, for their molecules actually attach themselves to individual fibres. The strength of this link may be increased by treating the fibres with a mordant (*eg* rosin and alum). The dyes used in papermaking are all synthetic, and fall into three main classes, direct, acid and basic. One of the drawbacks of these soluble colouring matters is that their lightfastness only ranges from poor to fair.

Optical bleaching agents are fluorescent dyes which are included in some papers, as well as in certain well-known brands of washing powder, in order to increase their brightness and whiteness. These remarkable materials are able to absorb invisible ultra-violet rays and re-emit them as visible light in the blue region of the spectrum. A paper can be made whiter by extending the bleaching process, but as well as adding to costs this is bound to cause some fibre damage. Papers can also be made whiter by adding blue dye to correct the yellow tint of the pulp, but this results in a reduction in the total amount of reflected light or in other words the production of an 'off-white'. The great attraction of optical bleaching agents is that they avoid these difficulties in giving a paper extra whiteness.

6. Paper and board – manufacture

PREPARING THE STOCK FOR PAPERMAKING

So far we have concerned ourselves with the raw materials for papermaking, the cellulose pulps obtained from wood and other plant sources, and the non-fibrous materials which are blended into paper. We now have to consider the methods of treating these raw materials in preparing the stock for the actual papermaking operation.

In this country these processes of stock preparation can be considered in two distinct stages, both involving a mechanical treatment in the presence of water:

(1) *Slushing, breaking or defibring* in which the wood pulp arriving at the paper mill in sheet form is broken up, to reconvert it into a suspension of the individual fibres in water.

(2) *Beating and refining* in which the fibres are physically modified so that in the subsequent papermaking operation they will come together to form a paper with the desired characteristics. In practice the non-fibrous ingredients of the paper are also added at this stage.

It is important to realise that the first of these stages effectively disappears in an integrated mill, combining both pulp and paper manufacture, since there is no need to produce the pulp in sheet form between the two operations. The trend towards integrated mills and the continuous processing of fibres is making several of the stages established for batch processing less distinct.

Slushing, breaking or defibring

This operation is commonly carried out in a *hydrapulper*, an open cup-shaped tank up to 6m in diameter, with an average depth of 3m (fig. 6.1). A high-speed propeller revolves in the bottom of the tank, throwing the pulp sheets and water up the sides and then into the swirling centre of the tank. Both the propeller and the lower walls of the tank are fitted with blades which assist the disintegration of the pulp. A perforated plate in the base of the tank allows fully disintegrated pulp to pass out into a storage chest. Hydrapulpers are used to break up waste paper as well as bales of pulp.

69

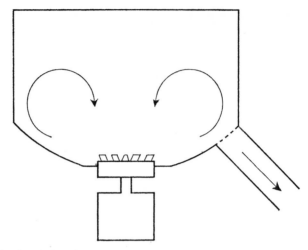

6.1 Principle of the hydrapulper.

Beating and refining

These processes form a key stage in papermaking for it is here that the final characteristics of a paper are determined. Paper made from an unbeaten pulp would be extremely weak and flabby, since there would be little adhesion between the fibres. In recent years most mills have supplemented or replaced the batch process of beating with the continuous process of refining. However, beaters are still in use, particularly in mills producing fine papers.

The basic principles of a beater have altered little since its introduction as the Hollander in the eighteenth century. It consists of an oval-shaped trough made of cast iron, wood, concrete or stainless steel, holding up to two tons of pulp (fig. 6.2). The stock is circulated round a vertical wall in the centre of the trough by the action of a heavy rotating roller, between the central wall and one side of the trough. The floor of the beater slopes up to a high point immediately in front of the beater roll. The stock is carried over this point and then falls down the other side so that its circulation round the beater is continuous. The mixture treated in a beater normally has a fibre concentration of about 5%.

The heavy beater roll has a series of metal bars arranged round its circumference projecting from the surface parallel to the shaft. The action of beating takes place between these roller bars, moving at about 600m/min, and the projecting bars of a bedplate set into the solid floor of the beater. The number, shape and spacing of the bars in both the roller and the bedplate depend on the nature of the pulp being beaten and the type of paper being made. The distance between the roller bars and the bedplate can be adjusted as beating proceeds, and it is here that the skill and judgement of the beaterman is exercised. By varying beating conditions it is possible to produce papers as different in character as blotting and greaseproof, starting from the same sulphite pulp.

This flexibility is possible because beating conditions can be adjusted to affect the fibres in different ways. For example, if the bars are of sharp design and close together they will tend to cut the fibres (fig. 6.3a). Resulting papers will be opaque, bulky and absorbent but not strong (*eg* blotting paper). If the bars are sharp but further apart there will be less cutting and a greater tendency to fray out or fibrillate the fibres (fig. 6.3b). The tiny fibrils protruding from

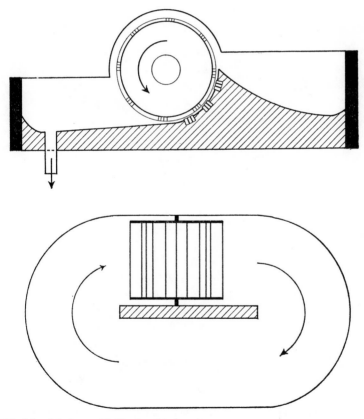

6.2 Principle of the beater (a) side view (b) plan view.

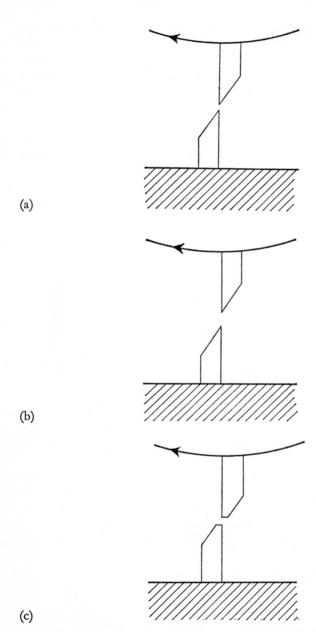

(a)

(b)

(c)

6.3 Varying beating conditions (a) bars sharp and close together. (b) bars sharp but separated. (c) bars blunt and close together.

the fibres help them to intermesh on the papermaking machine and so assist the formation of a strong sheet. If the bars are relatively blunt but close together, the fibres will be pounded together with water and will tend to lose their identity (fig. 6.3c). Eventually the fibres will be reduced to a 'cellulose jelly', and hard transparent papers like the glassines will result. This effect is known as 'hydration', since water is absorbed by the fibres during the process.

In practice these three effects of cutting, fibrillation and hydration go on together during beating, but conditions can be adjusted to favour one more than the others. It is important to realise that fibres from the various plant sources respond differently to beating. The short fibred esparto or soda wood pulps do not fibrillate or hydrate readily so they contribute bulk and opacity to a paper rather than strength. Rag pulps are far more responsive to beating and can be both easily fibrillated, and cut. Sulphite chemical wood pulps do not fibrillate particularly well but they can be hydrated or cut, and so are good general purpose fibres.

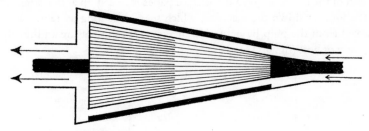

6.4 Principle of conical refiner.

The general effect of beating is to decrease opacity, bulk, tear strength, dimensional stability and the tendency to fluff, and to increase transparency, burst and tensile strength, density and hardness.

Refining was originally introduced to supplement and follow the process of beating but today it is often an alternative, particularly in preparing stock for newsprint and other high speed paper machines. The simplest form of refiner consists of a cone revolving rapidly inside a conical shell (fig. 6.4). Both the cone and the inside of the shell are fitted with bars, which can be compared to those fitted to the roll and bedplate of a beater. The pulp enters the narrow end of the cone, passes between the stationary blades of the shell and the rapidly rotating blades on the cone and is discharged at the wide end. As with a beater, the clearance between the two sets of bars can be adjusted to vary the nature and intensity of the beating action. Other types of refiners have been developed in which the beating action takes place between two discs, one stationary and the other rotating.

A single pass through a refiner may follow beating to complete stock preparation or alternatively batteries of refiners may be used to replace the beating process completely. With the development of modern refiners operating at high speeds and efficiency beaters are largely being replaced except in the production of fine papers.

The non-fibrous additives including fillers, sizing agents and colouring matters described in Chapter 5 were formerly always added to the beater. However, with the increasing use of refiners the trend is towards the continuous addition of these materials in metered quantities.

THE PAPER MACHINE

Although the modern paper machine is large, complex and expensive, its function can be summarised very simply as the removal of water from a dilute suspension of pulp so that the fibres come together to form a thin dry continuous web. The diluted pulp, containing about 99% water, flows on to one end of the machine and a dry web of paper is reeled up at the other. During the journey down the machine, which may be as long as 120m, water is removed from the pulp by natural drainage, suction, pressure and finally by the action of heat (fig. 6.5).

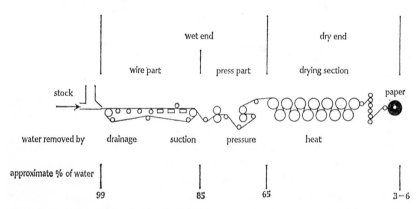

6.5 Main sections of a papermaking machine.

The general principles of a modern paper machine are not very different from those of the original Fourdrinier machine built in 1803. Diluted pulp flows on to a moving wire gauze, through which water can drain, assisted by suction. At the end of this section, the web of wet pulp is lifted from the wire on to a moving felt, which carries it between two or more pairs of squeezing rolls, so removing more water. Finally the web is carried round a series of heated drying cylinders before being reeled up. In this final section of the machine, the paper may be calendered to a smooth finish, surface sized or surface coated depending on the type of paper being produced. We will now consider these three sections of the machine, the wire part, the press part and the drying section in rather more detail.

The wire part

After the processes of beating and refining are complete, the stock is fed into the *machine chest*. Before flowing from this tank on to the wet end of the paper machine the stock must be diluted to the right consistency, brought to the correct rate of flow and finally cleaned. We saw earlier how foreign matter was removed from pulp during its manufacture from wood and other plant materials. Further cleaning is always necessary at this later stage, since beating and the addition of non-fibrous materials leave the stock containing particles which would cause trouble on the paper machine. The methods used to separate these unwanted materials are similar in principle to those already described for pulp manufacture, in that they rely on differences in density and size between acceptable and unacceptable particles.

From the strainers the stock flows into a *breast box* or head box, which maintains a constant head of stock above the wet end of the machine. At the base of the breast box is a slot (*the slice*), through which the pulp flows on to the wire. The weight per unit area or *substance* of the paper being made and the evenness of flow across the machine can be controlled by adjusting the slice.

If on hitting the moving wire, the fibres are subjected to a sudden acceleration, a high proportion of them will tend to line up in the *machine direction* and the properties of the resulting paper will be different in the machine direction from those in the *cross direction*. For example, tensile strength will be greater in the machine direction whilst tear strength will be higher in the cross direction. It is therefore desirable for the stock to be projected from the slice at approximately the speed of the wire. This part of the papermaking process has received considerable attention and has led to the development of more sophisticated devices for delivering stock on to the wire.

The wire itself is an endless belt of woven wire gauze, with a mesh varying from 12–40 wires per cm, depending on the type of paper being produced. Traditionally the gauze is made of phosphor bronze, but nylon has been used with considerable success. The wire is supported at each end by two large rollers, the *breast roll* near the slice and the *couch roll*, which drives the wire and brings it round on its return journey to the slice (fig. 6.6). Smaller rollers provide additional support between the breast and couch rolls, and others underneath keep the wire taut. The first part of a wire may be shaken from side to side in an attempt to counteract the tendency of fibres to line up in the machine direction. The shaking action is applied to the breast roll end of the wire with the other end being fixed. Thus *the shake* becomes progressively less as the stock moves along the wire. Since the fibres are arranged in their final positions when they have travelled about two thirds of the way down the wire, there is no point in continuing the shake after this stage.

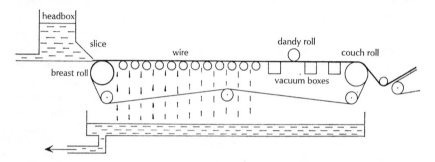

6.6 The wire part.

On some slower machines the stock is prevented from flowing over the sides of the wire by deckle straps, thick continuous rubber bands which move in contact with the edges of the first part of the wire. These are unnecessary on faster running machines. The width of paper made on a machine is called its deckle.

Towards the end of its journey on the wire, the wet pulp passes over several *suction boxes*. These rectangular boxes are connected to a vacuum pump, and draw water from the pulp through the openings on their surface. After the first or second suction box, the pulp passes under the *dandy roll*, a hollow wire-covered cylinder, whose function is to consolidate the sheet and apply a water-mark when required. The dandy roll is supported in such a way that it only rests lightly on the web, but applies sufficient pressure to impress the water-mark design into the paper when one is required.

The ease with which water drains from a pulp on the wire depends to a large extent on the way in which it was beaten. For example, a pulp which has been strongly hydrated in order to make a greaseproof or a glassine, will only drain slowly. On the other hand a pulp which has not been hydrated to any extent, *eg* mechanical wood for newsprint, chemical wood for blotting, will readily part with its water. This *freeness* or *wetness* of a pulp is an important

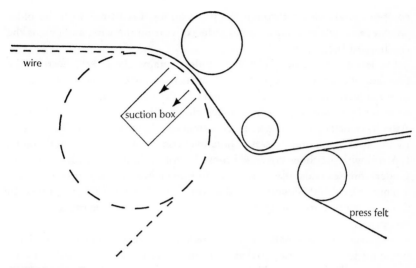

wire

suction box

press felt

6.7 Suction couch.

property for the papermaker since it influences many factors including the speed at which a machine can be run, the length of the wire and the number of suction boxes.

As the water drains through the wire it carries with it particles of filler and some of the smaller fibres ('fines'). As one might expect, more of these are lost from the wire side of the paper than from the topside, so that in talking of the 'two-sidedness of paper' we are not simply referring to the presence or absence of a wiremark but also to a difference in composition between the two sides. This leads to differences in the optical properties and the surface strength between the two sides.

The press part

As well as driving the wire, the couch removes more water from the wet web of pulp and provides the means of transferring that web from the wire to the first press felt. In the older type of couch used on slower machines, the wire carries the pulp between a felt covered top couch roll resting on a large plain brass roll which drives the wire. The pressure of the top couch and the absorbency of its felt removes water from the web so that it is sufficiently strong to make the short journey from the wire to the press felt. The modern type of couch relies on suction rather than pressure to remove water, and consists of a single perforated cylinder containing a suction box (fig. 6.7). The web may be transferred from the couch to the felt by means of a blast of compressed air blown through the perforations. Compared with the older *pressure couch*, the *suction couch* gives reduced wear on the wire, and avoids the crushing of bulky papers.

On leaving the wire the web of pulp contains about 85% water. The function of the press part of the machine is to remove as much of this water as possible from the sheet before it enters the drying section, since the removal of water by means of pressure is much cheaper than its evaporation by heat. After its transfer from the wire to a continuous felt, the web is carried through two to four squeezing presses. These may consist simply of a heavy granite top roll with a rubber covered bronze bottom roll, but as with the couch, modern presses make use of suction. In these newer *suction presses* both rolls are normally rubber covered but the bottom roll contains a vacuum box. In this way, more water can be extracted without increasing the pressure on the wet paper.

The last press is normally a *reversing press*, where the web travels through the nip in the reverse direction to that of the machine. The side of the paper which has previously been pressed in contact with the felt is here in contact with the roll surface. This arrangement helps to reduce the tendency of a paper to be two-sided, with the wire mark and the felt mark clearly impressed on each side.

Both the wire and the felts, which support the paper as it moves over the wet end of the machine, tend to become clogged with particles of fibre and rosin, etc. If they are allowed to build up these particles will prevent even drainage through the wire and reduce the absorbency of the felt, so both wire and felts are washed with water as they make their return journey down the machine.

The drying section

As the web leaves the presses it still contains about 65% water, so that for every ton of paper reeled up at the end of the machine about two tons of water have to be driven off in the final drying section. This formidable task is carried out by bringing the web in contact with a series of steam heated hollow cast-iron cylinders, about 1·5m in diameter. The number of cylinders depends on the nature of the paper being made and the speed of the machine. For example, a high speed newsprint machine may have as many as sixty drying cylinders. These cylinders are normally arranged in two layers. A dryer felt serving a group of several cylinders holds the web in close contact as it travels alternately under and over the successive cylinders, so that each side of the paper is dried in turn (fig. 6.8).

Steam is passed into each cylinder under pressure, and condensed water and unused steam are removed. It is important that the drying process is gradual, and cylinder temperatures are steadily increased to a maximum about three-quarters of the way down the drying section, after which they fall off slightly. To ensure uniform drying the polished cylinder surface is kept clean by scrapers or doctor blades.

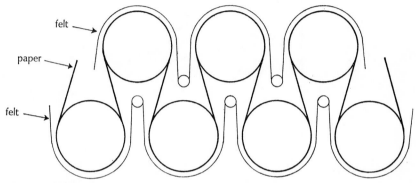

6.8 Part of drying section.

A number of factors influence the efficiency of the drying process. The cylinders must be arranged in such a way that there is room for moist air driven out of the paper to be quickly drawn away. Modern machines have hoods over the drying section to speed this withdrawal. The dryer felts must remain absorbent and on their return journey to pick up the web again they pass through small drying units.

The drying process has considerable influence on the final characteristics of a paper including such properties as its dimensional stability, surface finish and its tendency to pick. It is particularly important to maintain close control on the tension of the web during drying, so that the strains built into the paper are kept to a minimum. On modern presses, groups of cylinders have independent unit drives which can be automatically synchronised with one another to ensure tension control between one group of cylinders and the next.

About two-thirds of the way down the drying section the web may pass between a pair of heavy calender rolls, which improve the surface finish of the paper while it is still wet. This is also the point on many machines where the surface of the paper is treated with a solution of size and in some cases with mineral coating. In the past these were normally 'off-machine' processes, carried out as separate operations after papermaking. More recently economic pressures have led to a trend towards 'on-machine' sizing and coating.

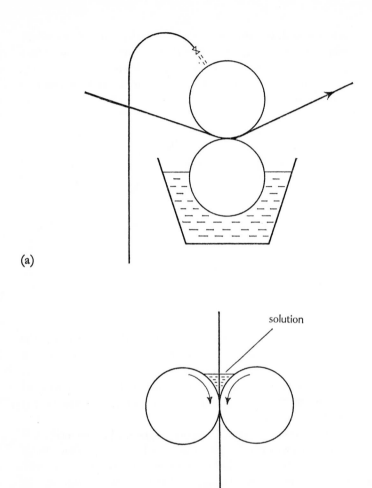

(a)

(b)

6.9 Size presses: (a) vertical; (b) horizontal.

Tub-sizing, the operation in which reels of paper are passed through a bath of size (normally a solution of gelatin, alum and formalin), is now only carried out in the manufacture of relatively few papers, *eg* high grade writings. On the other hand, *surface sizing* has become a standard practice on most paper and board machines. The most commonly used surface sizing solution is starch, but other suitable materials include carboxymethylcellulose (CMC), gelatin, polyvinyl alcohol and wax emulsions. In the vertical size press (fig. 6.9a) the web passes between two rollers, the lower one rotating in a solution of size and the upper one being fed with the same solution. In the horizontal size press (fig. 6.9b) the solution is held in the nip of two rollers placed side by side. Surface sizing has led to several improvements in the properties of paper, many of which are important from the printer's point of view. For example, it has greatly reduced fluffing, the tendency for loose fibre to come away from the paper surface. Until quite recently, this was a common problem on offset litho machines. Surface sizing has also improved smoothness, dimensional stability, water resistance and sheet strength.

At one time, the coating of paper with mineral pigment and binder was an off-machine process following papermaking. However, in recent years there has been an increasing demand for less expensive printing papers having some of the desirable characteristics of coated papers. This has led to the development of a variety of on-machine coating techniques and to the rapid growth in the use of these machine coated papers. The simplest method of *machine coating* makes use of the size press already described, but the various methods of coating paper whether on or off the machine are dealt with on page 92.

After leaving the last drying cylinder the web passes through one or more calender stacks, sets of heavy rollers, which give a further improvement in the finish. Where a particularly high finish is required water may be run into the nip of the calender. At this stage the paper is both hot and dry, and before finally being reeled up it may be passed over cylinders, through which cold water is circulating. As well as being cooled, the paper absorbs a film of moisture, which condenses on the cylinder from the atmosphere, and so is brought nearer to a point where it is in equilibrium with a normal atmosphere.

There are a number of possible variations from the basic Fourdrinier machine. On the *Machine-glazed (MG) or Yankee* type of machine much of the final drying of the web takes place against a large highly-polished cylinder, up to 5 m in diameter (fig. 6.10). A rubber-covered roll presses the semi-wet web against the steam heated cylinder, so that it is in close contact as the cylinder carries it round. In this way the paper is dried and at the same time glazed on one side. The characteristic MG finish can be seen on brown kraft wrappings, paper bags and poster paper.

Another variation on the Fourdrinier machine is to combine the wet webs from two or more wires to make one paper. By putting the two wire sides together, it is possible to avoid the two-sidedness which is an inevitable feature of papers made on a single wire. Alternatively the layers can be of different materials (duplex and triplex papers and boards) as, for example, in envelope paper having a coloured lining.

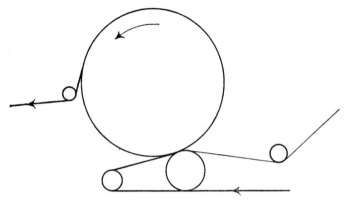

6.10 Paper passing round a machine glazing (MG) cylinder.

BOARD MANUFACTURE

When a paper exceeds a certain substance it is more correctly termed a board. In Britain, the dividing line is 220 g/m² and the term 'board' is taken to include a wide range of materials including boxboards, cardboards, mill-boards and strawboards.

Although we shall now be considering the special machines that have been developed to manufacture these heavier materials, it should be remembered that boards can also be produced on the single or multi-wire Fourdrinier machines described earlier in this chapter.

The *Cylinder machines* which now produce large quantities of board, particularly for packaging purposes, have their origin in the machine patented by John Dickinson in 1809, only six years after the building of the first wire machine for the Fourdrinier brothers. In the cylinder machine, a wire screen cylinder revolves in a vat containing diluted pulp. A layer of wet pulp is formed on the cylinder mesh as it moves through the pulp. On breaking the surface water drains through the mesh and then the pulp layer is transferred to the underside of a continuous felt, pressed against the top of the mesh cylinder by a rubber roll (fig. 6.11). The modern board machine may have as many as eight vats and cylinders each transferring a layer of pulp on to the underside of the same continuous felt (fig. 6.12). The wet layers of pulp knit together to form a single web of board. This system of manufacture has the advantage of great flexibility since as well as being able to increase the board thickness by varying the number of vats, it is also possible to feed different pulps to these vats and so produce boards with plies of different composition. For example, carton boards are normally made up of two or more different plies. One such board might contain:

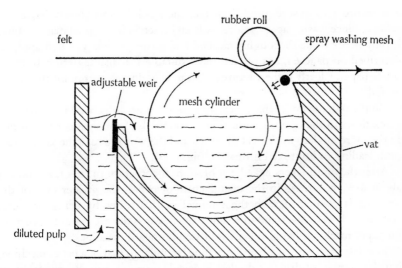

6.11 Contraflow vat.

(a) a top layer, consisting of a strong chemical pulp and some loading, providing a white smooth surface suitable for printing;
(b) a middle layer of poor strength and colour (grey), consisting of waste paper pulp;
(c) a bottom layer of only moderately good colour and strength containing a mixture of chemical and mechanical wood pulps.

Two or more vats normally contribute to each ply. In the example given above, vats 1 and 2 might form the top layer, vats 3, 4, 5 and 6 the middle layer and vats 7 and 8 the bottom layer, so that although the board produced may contain 50% waste paper it can nevertheless have a reasonably good appearance and printability.

The preparation of a pulp for board manufacture follows the same general principles that have already been outlined for paper manufacture. The greater thickness of board means that more water has to be removed on the machine

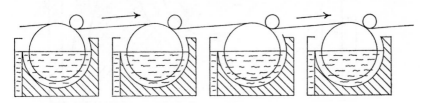

6.12 Part of a multi-vat board machine.

and in order to ensure good drainage, the stock is only lightly beaten or refined. Since waste paper has already been beaten in its original manufacture, it only needs to be thoroughly cleaned and disintegrated. The hydrapulper, described on page 69, has proved most suitable for disintegrating bales of waste paper. Clearly if the vats are going to contain different pulps they must be fed from separate head boxes.

On the older cylinder machines, *contraflow* vats are in general use. Here the stock flows in the opposite direction to the mesh of the cylinder (fig. 6.11). In the more modern *uniflow* vat the stock and the wire move in the same direction, leading to improved formation (fig. 6.13).

After the multi-layers have been brought together on the felt they contain about 85% water. As in the manufacture of paper, the greater part of the machine is concerned with the removal of this water by using pressure, suction and heat. The trend in recent years towards the greater use of suction on paper machines has also been followed on board machines.

The drying section of a board machine is similar to that of a paper machine, but the number of drying cylinders is usually much greater. Board machines are generally slower since larger quantities of water have to be removed. The drying section contains a calender stack or stacks and may include a surface sizing or surface coating unit. On some board machines the reel is finally slit and cut into sheets.

The *Inverform process* introduced in 1958 was an important development in

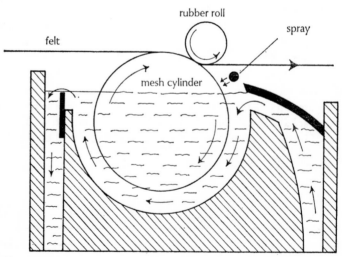

6.13 Uniflow vat.

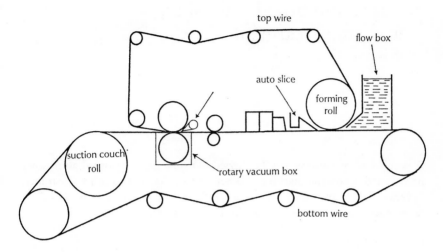

6.14 Principle of the Inverform process.

the manufacture of boards and heavy papers. Whilst on a normal wire machine the removal of water by drainage and suction takes place downwards, on an Inverform machine, water is removed in both a downward and an upward direction (fig. 6.14). Shortly after leaving the flowbox the stock is sandwiched between a bottom wire and a top wire under a forming roll. The two wires, with stock held between them pass through the *auto slice*, a stiff scraper blade held lightly inside the top wire in such a way that water is forced out of the stock, up the sloping blade into a tray, from which it is led away. More water is extracted from the stock by suction boxes placed under the bottom wire and over the top wire. Several of these Inverform units can be combined to produce boards or thick papers made up of a number of plies. After laying the first ply, most of the water from subsequent layers is removed upwards through the top wire. The main advantage of Inverform machines is that they are able to operate at higher speeds than traditional machines, but it is claimed that the process also gives improved formation and ply adhesion.

One other process of board manufacture must be mentioned, that is the much older method of producing *Millboards*, the strong rigid sheets used in bookbinding as case boards. A wet layer of pulp normally from a cylinder unit is wound on to a wooden or cast-iron making roll, until it has built up to the required substance. A divider cuts the sheet, which is then pressed between cotton or jute sheets. After drying in a tunnel the sheet is conditioned and finally 'milled' between heavy rollers to give a hard smooth finish.

7. Paper and board – finishing and after treatment

A number of operations may be carried out on a paper or board after it has been reeled up on the machine and before it is despatched to the customer. Several of these processes influence the 'finish' or surface properties of the paper and so are of great importance to the printer.

CALENDERING

This process of pressing a paper or board between heavy rollers clearly has a large effect on surface characteristics, particularly smoothness and gloss. When only a low or medium finish is required this can be achieved by calendering on the paper machine, but if a high finish is desired then the paper is supercalendered as a separate operation after papermaking. Whilst machine calenders primarily rely on pressure to achieve their effect, supercalenders make use of both pressure and friction.

On the paper machine itself, the number and type of calender stacks can be varied to obtain a range of finishes (machine finish or MF). A stack of intermediate calenders may be included before the last section of drying cylinders (page 79) and depending on the finish required, there may be up to five stacks of machine calenders at the end of the drying section. Each stack consists of three, five or seven polished chilled iron rolls, some of which may be steam heated. An imitation art finish may be obtained on the paper machine by applying a thin film of water to the paper as it enters the nip of one or more of the calender units. Where this is done the paper must then be passed through at least two further calender stacks of heated rolls. While discussing finishes obtainable on the paper machine alone, reference must be made to the machine-glazed (MG) papers produced on the special machines described on page 81. This single-sided glazed finish is achieved by drying the semi-wet web in close contact with a large highly polished heated cylinder.

If a machine finish is inadequate the paper is damped by a fine spray of water just before being reeled up on the paper machine and later is supercalendered on a separate unit. A supercalender consists of a vertical stack of an even number of rolls. This number is normally between eight and twelve but for glassine may be as high as eighteen. Unlike the machine calenders where all the rolls are of chilled iron, supercalenders have more resilient fibre rolls (bowls) alternating with slightly smaller steam heated iron rolls. The fibre bowls are made by threading rings of cotton, woollen or linen paper on to the shaft of the roll, compressing them so that their edges form the roller surface and finally polishing this surface.

The top and bottom rolls of the stack are of chilled iron and these alternate both up and down the stack with the fibre bowls. This arrangement brings the two fibre bowls together in the centre of the stack, so that the surface of the paper that has been in contact with the steam heated iron rolls in the first

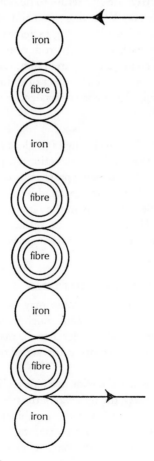

7.1 Principle of the supercalender.

half of the supercalender is brought against the fibre bowls in the second part of the stack and vice versa (fig. 7.1.)

Supercalenders normally operate at linear pressures ranging from 35kN/m to 250kN/m (glassine 500kN/m), but in addition to this nip pressure, the slipping action of the resilient fibre rolls makes a major contribution to the finish obtained, since it has the effect of polishing the paper. This polish depends on the combined effects of pressure and speed of travel, both of which can be varied. A considerable degree of skill is required in matching these variables for a particular paper. Papers made from well beaten fibres will respond more to calendering than bulky resilient paper in which the fibres have retained their original structure and refuse to lie down.

On earlier pages in this book blotting paper and glassine have been given as examples of familiar materials illustrating extremes in the paper making process. In the first of these one is seeking to produce an open absorbent structure, in the second a smooth compact transparent film. Clearly blotting paper is not calendered, since this would have the effect of reducing its bulk and absorbency. On the other hand, glassine is supercalendered under the highest pressures in an eighteen roll stack.

Friction glazing is a method of producing high gloss paper for wrapping highly polished articles like cutlery and for confectionery, particularly bars of chocolate. The paper is damped before being passed through a calender in which two small heated waxed iron bowls are separated by a paper bowl. Neighbouring rollers are driven at different speeds in the ratio 2:1 and the combined effect of friction, pressure and heat on the waxed surface is to produce a glossy skin on the paper. When coated papers are friction glazed it is possible to include the wax in the coating mix.

SLITTING, TRIMMING AND CUTTING

A reel of paper from the paper machine may be anything from 1500mm to 850mm wide. Certainly it is normally much wider than the reels or sheets required by the customer and also wider than the machines on which finishing operations like calendering and coating are carried out.

Slitting is the process in which the wide machine reel is sliced into smaller reels. It is combined with *trimming*, the removal of the two rough deckle edges from either side of the machine reel. On the combined slitter, trimmer and re-reeler the machine reel passes over a cutting table at high speed and under controlled tension. Slitting and trimming takes place between pairs of adjustable circular knives, an upper knife being set at an acute angle to a lower knife which is driven by a rotating shaft. The resulting smaller reels are wound on to cardboard centres, each cut to the correct width and mounted together on the same spindle, from which they can later be removed. The shavings of deckle edge trimmed from each side of the reel are repulped.

Cutting the reels into sheets may be carried out as a separate operation from slitting, but if the slit reels are not going to undergo any other finishing process, the slitting and cutting operations can be combined. On these combined units, re-reeling is unnecessary and the slit reels pass between rollers on to the cutting stage.

In the cutting process the web passes over a fixed knife (*dead knife*) running across the width of the cutter, and the cut is made by the shearing action of a *chop knife* mounted at an angle to the dead knife on a rotating drum. The length of sheet cut is adjusted by varying the speed of the rotating drum.

Modern high-speed cutters are usually fed with slit reels rather than the original machine reels. They may be linked with as many as twelve reels at a time, so that each cut of the chop knife produces twelve sheets. While cutting is taking place, a further twelve reels can be mounted on a turntable device, so that when the first reels are empty, a new stand can be swung round in line with the cutters, and the 'down-time' of the machine kept to a minimum. This method of cutting several reels together is of considerable significance for the printer since it means that successive sheets being fed through a printing machine can come from different reels. Normally one would expect the paper in a set of reels to be well matched, but there have been instances of printing faults recurring at fixed intervals in a pile of paper. If, for example, one reel in a set of four feeding a cutter had a much lower oil absorbency than the other reels, the printer could find that every fourth sheet in the stack held more ink on the surface of the paper. This could result in the print having a different visual appearance and a different rate of drying.

Refinements that have been added to modern cutters include automatic counting of the sheets, devices to remove static electricity and the automatic removal of piles of sheets on a conveyor when they reach a preset height or number of sheets.

It is vitally important for the papermaker to ensure that the reel and sheet sizes he is supplying to his customer can be obtained economically from the full width (deckle) of the paper machine, without excessive cutting waste. In order to do this a technique known as *duplex cutting* may have to be used. If, for example, the machine reels were 2600mm wide after trimming and the paper is being ordered in sheets 768mm × 1008mm (metric quad crown) the reel could be slit into two reels 768mm wide and one reel 1008mm wide. For the two 768mm reels, the speed of the chop knives could be set to give sheets 1008mm long and for the 1008mm wide reel it could be adjusted to give a sheet 768mm long. Although in this case the method has enabled the full width of the machine to be used in producing metric quad crown sheets with no waste, the result may not be acceptable to the printer. Whereas the sheets from the 768mm reels have their long edge in the machine direction (long grain), those from the 1008mm reel have their short edge in this direction (short grain). When the moisture content of a paper changes, the expansion or contraction in its size is much greater in the cross direction than in the machine direction. In order to keep these changes in size to a minimum, printing papers are normally cut long grain. When several colours have to be printed, any change in paper size between separate printings can cause misregister. The greatest chance of this happening is in litho printing when the water on the press causes paper to pick up about 0·2% of water on one pass through the machine. On many machines the printer is able to adjust print length but not print width; a long grain sheet is therefore advantageous here, in that the greatest movement of the paper corresponds with the ability to adjust print length (fig. 7.2).

If duplex cutting cannot be used in the example given above, the paper-maker could either decide to produce three 768mm reels and one small 296mm reel or he could move the deckle straps on the machine wire so that a 2304mm trimmed reel was produced. In this sort of situation the papermaker's decision will to a large extent be influenced by external factors.

CONDITIONING

The dimensional instability of paper with changing moisture content is one of its greatest drawbacks as a substrate for printing. Changes in the size of a sheet between successive colour printings will cause misregister and because the superimposed colours do not fit properly together, the final print will lack clarity and crispness. When moisture is lost or gained by only part of a sheet, the changes in size are also local and the sheet loses its flatness. For example, if a stack of paper is moved into a humid atmosphere, the edges of the sheets will take up moisture faster than their centres and their expansion will lead to wavy edges. Wavy edges, tight edges and curl can all create feeding and delivery problems and cause the creasing of sheets on the printing ma-chine. As faster running machines are developed, paper flatness becomes absolutely essential to high productivity.

The moisture content of most printing papers as they come off the paper machine is between 3% and 6%. If this paper is taken into the average printing machine room it will absorb moisture from the air until its moisture content

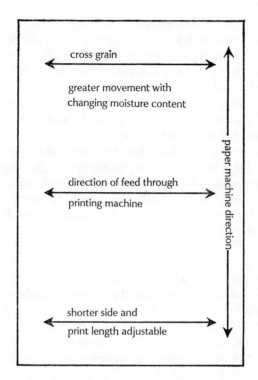

cross grain

greater movement with
changing moisture content

direction of feed through

printing machine

paper machine direction

shorter side and

print length adjustable

7.2 Advantages in printing a long grain sheet.

is between 6·5% and 8·5%, depending on the relative humidity of the press-room. In doing so it will increase in size in all three dimensions, the highest percentage increase being in its thickness and the lowest in the machine direction. The percentage increase in size in the cross direction may be as much as ten times that in the machine direction.

The objectives in conditioning are to add moisture to a paper so that any expansion takes place before it is printed or converted and to relieve the internal strains built into the paper when it was dried under tension on the paper machine. On modern conditioning machines the web is passed verti-cally up and down through a large number of narrow boxes. Inside these boxes it meets a current of air flowing in the opposite direction. The relative humidi-ty and temperature of this air is adjusted to an appropriate level by passing it through a system of water sprays and a refrigeration unit. The paper is finally re-reeled without tension. The moisture content of the final reel can be controlled by varying the relative humidity of the circulated air. If the printer is able to specify the temperature and humidity in his machine room, paper can be produced with a moisture content which will be in equilibrium with these conditions. If such an arrangement is made, then obviously close attention must also be paid to the conditions under which the paper is packed, transported and stored.

Another advantage of conditioning is that it removes the static charges which may have built up on the paper at the dry end of the paper machine, and by increasing the moisture content reduces the likelihood of static recurring in subsequent processes. Some printers still follow the traditional practice of conditioning or 'maturing' paper, by hanging sheets of paper in wads on a conveyor which circulates them in the machine room atmosphere for about 24 hours. Effective conditioning by the paper maker before the paper is cut into sheets should make this process superfluous, unless there are special circumstances.

COATING

The printing properties of a paper or board can be greatly improved by the application of a thin layer of fine mineral pigment and binder since the surface of a paper made from natural cellulosic fibres can never be completely uniform. The fine particles of mineral pigment in the coating mixture are able to cover many of the imperfections to give a paper with improved smoothness, ink receptivity and visual appearance, properties which are essential to good quality in the printing of fine halftones.

The coating mixture consists basically of two parts, a fine mineral pigment and an adhesive in which this is dispersed. The pigments used are chemically similar to the fillers or loadings added to paper itself (page 66) but they must be in a more finely divided form. Commonly used materials are specially refined grades of china clay, calcium carbonate, barium sulphate (blanc fixe), titanium dioxide and a mixture of calcium sulphate and alumina (satin white). The more expensive pigments, like titanium dioxide, are not normally used on their own but are blended in with other pigments. This mixture of pigments is thoroughly dispersed in the binder or adhesive, whose function is first to act as a vehicle on the coating machine and then to hold the layer of pigment firmly to the paper surface. The adhesives in coating formulations include natural materials like starches and casein, and synthetic polymers like styrene-butadiene and butadiene-acrylonitrile. Normally these synthetic polymers are used in the form of aqueous dispersions known as latices, and these are combined with casein or a modified starch in the coating mixture. A typical coating mixture has a milky consistency and might contain one part of adhesive to seven parts of pigment. Other additives which may be present include a defoaming agent, dyes or pigments to colour the coating, formaldehyde as a hardener and a wax emulsion to improve the finish.

Although a layer of coating can to some extent disguise some of the shortcomings of the paper underneath, not all papers are equally suitable for coating. Suitable characteristics for coating must be planned into a good base paper. In particular it should be flat, reasonably smooth and free from surface faults which can be magnified by the coating. Above all, the paper must be uniform in its substance and texture, otherwise the coating and any subsequent printing will show similar irregularities. The extent to which the smoothness of a coated paper depends on the smoothness of the base paper varies with the coating method. For instance, in the air knife process the coating is of fairly uniform thickness and so tends to follow the 'hills and valleys' of the base paper.

Paper coating methods have passed through an extremely active period of development in recent times. Traditionally coating was carried out as a separate operation after papermaking (off-machine), but today large quantities of paper and board are coated on the paper machine (on-machine, machine-coated). The classical method of brush coating developed from the application of the coating by hand with a brush. Although brush coating is still used to produce some of the highest quality art papers, a great variety of other techniques for applying the layer of coating have been introduced, some of which can be used either on or off the paper machine. These newer methods have made it possible to produce paper and boards with good printing quality at high speed and low cost.

Methods of applying coating can be considered under the following headings: Brush; Air knife; Roll; Metering bar; Blade; Cast.

In *brush coating* the web of base paper passes round a large drum, and coating mix is applied to its exposed surface by means of a rotating brush or roller

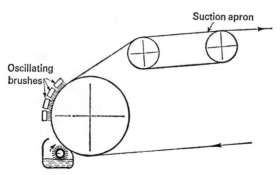

7.3 Brush coating.

(fig. 7.3). As the paper moves round the cylinder, the coating is smoothed out by a series of brushes, moving from side to side, with bristles which become successively finer. The amount of coating applied is controlled by the position of the pick-up roll in the coating mix and by the pressure of the smoothing brushes. After leaving the drum the paper may be drawn over a suction box and festooned on wooden sticks which carry the paper in loops through a heated drying chamber. The traditional brush coating machine with festoon drying only runs at about 60 m/min. When the coating is applied by a roll and festoon drying is replaced by the more modern tunnel drying, speeds may be increased to 280 m/min. Although the process is slow by modern standards, brush coating is still used to produce high quality coated papers.

In the *air knife* method, a roller applies an excess of coating mix, which is then metered and smoothed out by a controlled jet of air (fig. 7.4). As with brush coating the process is limited by the fact that the coating mix must have a low viscosity, from which it follows that the solids content is relatively low (35–45%) and a large amount of water has to be removed in the drying tunnel. Running speeds are also restricted by the tendency for a clay coating to mist at the air knife. Another characteristic of the process is that similar weights of coating are applied to both the 'hills' and the 'valleys' so that the contours of the base paper tend to be followed by the coated paper.

A great variety of *roll coaters* have been developed ranging from the simple size press to far more sophisticated systems. In the vertical size press (page 80) the web passes through the nip of two rollers which forms a small reservoir for the solution being applied. Size presses are commonly used on paper and board machines to apply a solution of size to both sides of the web,

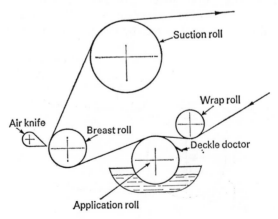

7.4 Air knife coater.

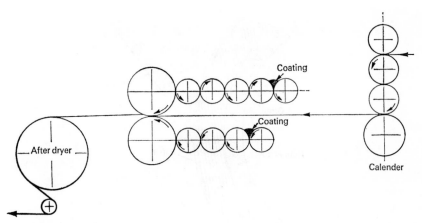

7.5 On-machine roll coater.

but they are also capable of applying a thin layer of coating in exactly the same way. The maximum weight of coating that can be applied by these units is low, but running speeds of 300m/min. are quite common. Other systems of roll coating have been developed with more efficient means of metering and spreading the coating mix. In the Massey process of on-machine coating, trains of small oscillating rolls deliver an even film of coating to each of two large applicator rolls driven at different speeds (fig. 7.5). These more sophisticated roll coaters have been designed to try to overcome the problem of the 'splitting' of the coating mix in the roller nip and the resulting 'orange peel' patterning on the coated surface.

In the Champion coating process, excess coating is applied by a reverse running roll and metered by a simple *metering-bar* (fig. 7.6). This bar may be replaced by a driven rod which may be smooth or wire wound. The method is simple, effective and widely used for on-machine coating at speeds up to 360m/min.

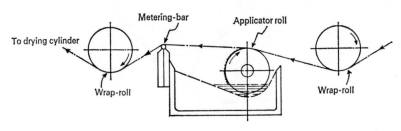

7.6 Metering-bar coater.

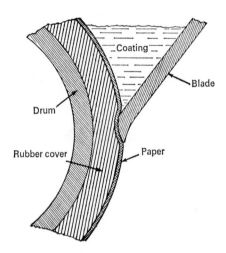

7.7 Blade coater.

Blade coating has been the most recent major development and is tending to re-place roll coating. Running speeds can be as high as 1200m/min. and the methods may be used either on or off the paper machine. In the simple blade coater, the web of paper passes round the surface of a large rubber-covered roller and is held against it by a flexible blade of spring steel (fig. 7.7). The

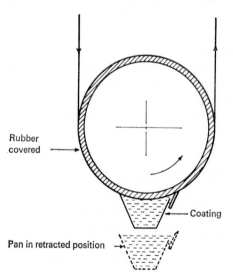

7.8 Flexiblade coater.

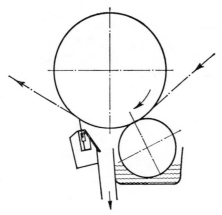

7.9 Inverted blade coater.

coating mix is held in the reservoir formed between the blade, the paper held against the roller and dams fitted at each end of the nip. More recently developed blade coaters are variations of this basic theme. In the *flexiblade coater* the coating mix is held in a trough beneath the cylinder, the blade forms the front of the trough and the system is enclosed so that the mix can be fed under pressure (fig. 7.8). This pressure provides convenient method of controlling the pressure of the blade against the web and also reduces blade wear. In the *inverted blade coater*, an excess of coating is applied by roller and a flexible blade wipes off the excess leaving a smooth coating (fig. 7.9). The excess coating is thus doctored off and re-used. All these blade coating devices allow the use of very high solids coating formulations (up to 65%), so that correspondingly high running speeds are possible. Blade coating methods generally produce a very smooth surface, free from the contouring effects associated with the air knife and the 'orange peel' pattern from roll coaters. Their disadvantages are that heavy coating weights cannot be satisfactorily applied and that any imperfections in the base paper are liable to cause web breaks.

Most of the coating methods already outlined may be used either on or off the machine. Large amounts of board are coated on-machine by the Champion and air knife processes. On-machine papers are generally the large tonnage grades like magazine papers and here roll coating and the Champion process have been widely used. Blade coating is being increasingly used and appears to have a great future potential. Off-machine coating methods have made a significant advance in recent years and faster running speeds and lower costs combined with their flexibility and quality continue to make them the most suitable methods for many papers. Two or more of the different coating

97

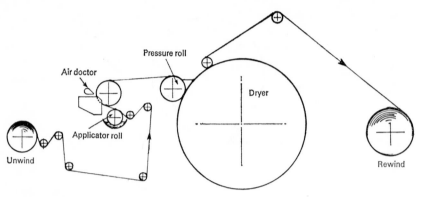

7.10 Cast coating.

methods available may be used together to combine the advantages of each. For example an air knife coat may be put on top of a blade or metering bar coating.

If a smooth finish is required then coating is followed by supercalendering or glazing. One exception to this is the very specialised process of cast coating, which produces the glossiest of all coated surfaces. The principle of cast coating is to bring the wet coating applied to one side of a paper in close contact with the highly polished surface of a heated drum (fig. 7.10). In this way the mirrorlike finish of the chromium plated drum is transferred to the coated surface of the paper as it dries in contact.

OTHER FINISHING PROCESSES

A number of other processes may be carried out on a paper after it is made, but generally these are of less significance to the printer than those already described.

Tub-sizing is the traditional process in which the web of paper is passed through a bath of size after it has left the paper machine. The bath normally contains a 5–7·5% solution of gelatin together with alum as a hardener and formalin as a preservative. On emerging from the solution, the excess of size is removed by rubber squeezing rolls and the web is carried over open cylinders with wooden slats to be slowly dried in warm air. Tub-sizing with gelatin is carried out on high grade notepapers with a rag content. Cheaper grades are normally surface sized on the paper machine.

In contrast to the traditional process of tub-sizing, there has been a considerable development in recent years in the coating or lamination of plastics on to paper and board. The resulting materials have become increasingly important for packaging applications where they combine the advantages of the paper or board, *eg* good printability, low cost, with the advantages of particular synthetic polymers, *eg* chemical resistance, flexibility, gloss, low moisture permeability. The most widely used of these polymers has been polythene, applied by the process known as extrusion lamination. Here granules of polythene are melted in a heated barrel and extruded through a die as a molten film directly on to the web of paper or board. A great variety of other polymers may be combined with paper or board either before or after printing.

The water resistance of paper and board for packaging applications can be improved by the process of *waxing*. Printing is normally completed before surface waxing is carried out, a familiar example of this being the production of bread wrappings. However, in another process, the wax is made to completely penetrate the body of the paper, and the surface is still able to accept printing ink.

Another wrapping material, *vegetable parchment*, is produced by passing a web of an absorbent paper through a bath of concentrated sulphuric acid. This reduces the fibrous structure of the paper to a gelatinous state, and after washing and drying a paper is obtained which combines chemical purity with greaseproofness and wet strength and so is an ideal material for wrapping fatty substances.

8. Paper and board – classes

Although paper is traditionally thought of as a material to write on and read from, its role in the modern world is far wider. As well as being the main carrier of written and printed words and pictures, paper and paper products find many other applications, notably in the packaging and building industries. The paper industry produces a great variety of materials including such things as soft tissues, creped towelling, insulating boards, fibre pipes, multiwall sacks and fibre drums. No attempt will be made in this book to consider all these products, and in discussing the various classes of paper and board the main emphasis will be placed on those materials which may be handled by the printer.

Previous chapters describing raw materials and manufacturing methods for paper have shown the many ways in which the papermaker can vary the properties of the paper finally produced. In the first place he has a choice of fibrous and non-fibrous raw materials; he may vary the amount of filler and the extent of sizing; beating conditions may be varied; the paper or board may be produced on different types of machine, and on any one machine, making conditions can be adjusted in many ways; finally the properties of the paper may be modified by coating, calendering or by some other finishing operation. With this flexibility, it is possible to produce an infinite number of papers, each paper having a particular set of properties, which can to a large extent be defined by the measurement of such things as thickness, bulk, smoothness, absorbency, gloss and opacity. In practice, for economic reasons the number of papers produced is much less than infinite and any one mill tends to concentrate on a limited range of papers. Nevertheless, even allowing for specialisation and standardisation, many hundreds of different papers and boards are available, and if the order is large enough and the price economic, a paper can still be made to meet the particular requirements of a customer.

The many different types of paper can be put into classes according to their applications, *eg* printing, writing. Although boards can simply be thought of as thick papers or combinations of two or more papers, they are generally considered as a separate group. The main classes can be considered under the following headings: Printings; Writings; Wrappings; Boards; Speciality papers. Within each of these headings there are great variations in quality and price.

PRINTINGS

All papers for printing must be receptive to ink and have reasonable strength, opacity and colour. A certain minimum strength is required for the actual printing operation, but beyond that come the varying demands made on the printed product during its lifetime. For example, although strength and durability are not important in the short life of a newspaper, they are essential for the pages of a reference book. If a paper lacks good opacity, printed matter shows through on to the opposite side of the sheet (show-through). The effect is particularly objectionable when the paper is printed on both sides.

Newsprint only barely meets the minimum requirements for a printing paper. It consists largely of mechanical wood pulp, but it may be strengthened with a small proportion of chemical wood. The high percentage of mechanical wood means that the paper has poor strength and durability, yellowing on exposure to light. Newsprint is not normally sized but it contains a reasonable amount of loading to reduce show-through. When supercalendered it is capable of reproducing coarse screen halftone illustrations. High oil absorbency is an essential feature of a newsprint, since this facilitates quick drying by the absorption of news ink on the high-speed presses.

As the proportion of chemical to mechanical wood pulp is increased the quality of a paper is improved and its cost rises. *Mechanical printings* are a range of papers in which the proportion of chemical wood is greater than that in newsprint. They are widely used for the cheaper kinds of publication work. In *woodfree or pure printings*, mechanical wood is omitted altogether and quality is further improved. *Esparto printings* in which esparto pulp is blended with chemical wood pulp are used for better quality publication work. The esparto content may be replaced by hardwood but in either case careful beating gives a high quality paper, with a characteristic softness, bulk and opacity. The term *antique* is used to describe a bulky paper with a rough finish.

The engine sizing of a paper primarily controls its water absorbency and writing papers must be so treated to prevent water-based inks from feathering out from the nib. This type of sizing appears to have a negligible effect on oil absorbency and so it is not strictly essential where a paper is to be printed with oil-based inks. Clearly, paper which is likely to be written on after printing (*eg* forms, magazines containing reply coupons, etc) should be engine sized. In practice most printing papers are sized to some extent and mechanical, woodfree and esparto printings are available in grades ranging from soft to hard sized. The choice of a particular grade will depend on the printing process being used and on the function of the printed product. Litho papers are generally medium to hard sized, gravure papers may be unsized or only soft sized, and papers likely to be varnished after printing should be hard sized. The surface sizing of an uncoated paper improves its varnishability as well as its resistance to fluffing (page 66).

Grades of mechanical, woodfree and esparto printings are also produced with different surface finishes ranging from the rough antique finish to the fairly rough *machine finish* (*MF*) and the smoother *supercalendered* (*SC*) finish. A machine finish is perfectly satisfactory for a book which consists largely of text and a few line illustrations. The book will have bulk, a property which publishers think is important at the point of sale. It will also be easy to read by artificial light, because of the absence of reflections from the surface, which can be distracting on a glossy paper. Esparto or woodfree *offset cartridge*, another paper with a relatively rough surface, is widely used for books and brochures printed by litho. However, when fine screen halftone illustrations are being printed, particularly by letterpress, the paper must have a smoother surface, in order to allow the smallest dots to be effectively transferred from the printing plate. Supercalendering provides an economical way of achieving a reasonably smooth surface on an uncoated paper and many S C grades of mechanical and woodfree printings are available. *Imitation art* papers, as their name implies, are an attempt to build some of the advantages of a coated art paper into an uncoated paper. These papers contain a high percentage of loading and are given a high finish by spraying the web with a fine mist of water just prior to calendering (water-finish). A smooth surface is also a requirement for *poster papers*, which are produced on special machine glazing (MG) paper machines (page 81). One side of an M G poster paper has a smooth glossy finish suitable for halftone colour printing by offset litho and the other side is rough so that the paper can be easily pasted to a hoarding.

A very smooth surface is required for high quality colour printing. There is a limit to the finish that can be achieved by supercalendering alone, but the prior application of a thin layer of mineral coating can produce great improvements in smoothness and general printability. A wide range of coated papers and boards are available. The coating may be applied as a separate operation or it may be applied on the paper machine itself. The various methods of coating are described on page 92. The quality of a coated paper depends on the method used but also on the quality of the body paper receiving the coating.

Art papers or boards are those coated as a separate operation, normally on to an esparto or chemical wood body. *Chromo papers* are generally only coated on one side. Proofing chromos are usually of a heavier substance than litho or label chromos. *Cast coated papers* have a particularly high gloss finish and are produced by a special technique (page 98). The increasing use of colour in advertising, packaging and printing generally has brought an insistent demand for papers with good printing quality but low cost. The wide range of *machine coated papers* now available on mechanical or woodfree bodies, have gone a long way to meet this demand.

WRITINGS

Like printings, writing papers must have good strength, colour and surface finish. They must have a consistently good appearance and be sufficiently sized to prevent water-based ink from feathering. Writings form a very large group of papers, in which there are great variations in quality, dependent largely on the choice of fibrous raw materials. For example, the highest quality writing, drawing and blotting papers are made from rag. On the other hand, the lowest quality writing papers contain a high proportion of mechanical wood, and could be described as engine sized newsprint. The many grades intermediate in quality are based on chemical wood with or without the addition of esparto or straw. Writing papers are usually either laid or wove, depending on the impression made by the dandy roll on the paper machine (page 76). The difference between the parallel laid lines and the wove can be seen by holding a sheet of paper up to the light. High quality writings containing rag are normally tub-sized.

Banks and *Bonds* are low substance writings produced in many different grades for office stationery of all kinds. They are normally based on chemical wood but the best qualities contain rag and are tub-sized. They are often produced with a parchment-like finish. The difference between banks and bonds is simply one of substance, banks being below $61g/m^2$ and bonds being above this substance. *Duplicator papers* are wood or esparto writings with the degree of sizing reduced so that they are sufficiently absorbent to prevent freshly printed ink from smearing. Although banks, bonds and duplicator papers are classified as writings, they are often handled by the letterpress or litho printer, particularly in the production of business stationery.

The very highest qualities of *drawing paper* are made from rag, tub-sized and air dried. Less expensive grades contain chemical wood pulp or esparto, and these drawing cartridges are similar to the popular offset cartridges used extensively in litho printing. In producing *blotting paper* the papermaker's objective is an open absorbent sheet, and this is achieved by creating large numbers of clear narrow channels capable of drawing in liquids by capillary action. Beating is restricted to a cutting action, and no size or filler is added since this would have the effect of blocking the channels.

WRAPPINGS

Wrapping papers are primarily required to give protection. Where strength is important the fibres should be long and beaten in such a way that they form a strong intermeshed structure.

In producing the familiar brown *kraft* wrapping papers, bleaching is omitted since this tends to reduce strength. The strongest kraft papers contain pure sulphate pulp. Other wrapping papers are based on sulphite chemical wood or on mixtures of sulphate and sulphite. They are produced in various colours and may be machine glazed on one side (page 81). These M G papers are widely used in making paper bags. As with M G poster paper, their glazed surface provides a reasonable surface for printing and the rough underside provides a good key for adhesives. Cheaper wrappings like *M G cap* contain mechanical wood pulp. The familiar buff coloured *manilla* papers and boards are generally made from unbleached sulphite and not from manilla hemp as was originally the case. *Lightweight tissues* are made from chemical woodpulp, and various grades are available. Although very different in appearance from these papers, *greaseproof* is also based on chemical wood. Its characteristics are developed at the beating stage, when the fibres are pounded to a hydrated jelly (page 73). If a good greaseproof paper is damped and heavily supercalendered the result is an almost transparent paper known as *glassine*. *Vegetable parchments* which resist water as well as grease are produced in a special process in which an unsized sulphite paper is given a rapid immersion in a bath of concentrated sulphuric acid.

BOARDS

When the substance of a paper exceeds $220\,\mathrm{g/m^2}$ it should strictly be considered a board. This completely arbitrary line of demarcation is laid down in British Paper Trade Customs and used in preparing statistics and for assessing such things as import duty. However, in considering the nature of board this dividing line between paper and board has little significance and boards can simply be considered as high substance materials which are essentially thick papers or combinations of two or more papers. Where the board is built up from several layers of paper, these layers may be of different composition.

The cheapest of all these materials are the *strawboards*, which are made on fourdrinier type machines from straw pulp in substances ranging from $225\,\mathrm{g/m^2}$

to 1000g/m². Large quantities of these strawboards are produced in Holland. They combine rigidity with low cost, and so find extensive use in making rigid boxes and tubes, in case and library binding and in making showcards. Substances over 1000g/m² may be made by pasting together two or more lighter boards. Another cheap range of materials, the *chipboards* are made entirely from wastepaper pulps. Although of similar quality to the strawboards, they are less rigid but have better folding properties. *Brown woodpulp boards* produced mainly in Scandinavia are based on mechanical wood pulp. They are available in several grades including unglazed, semi-sized and glazed, and they are used for a variety of purposes, *eg* showcards, boxes for flowers, clothing and machine parts. Better qualities of woodpulp boards contain a proportion of sulphite pulp. *White pulp boards* may have mechanical, woodfree or esparto furnishes and only differ in substance from the range of printings described on page 101. They may be glazed or unglazed and are produced in a variety of tinted shades.

Millboards are built up on the surface of a cylinder mould by allowing a layer of wet pulp to wind round it several times. When the correct number of plies has been brought together, the web of pulp feeding the cylinder is cut, and the sheet transferred to a table in front of the press. After pressing the boards are dried and then milled between rollers. Although the process is intermittent, slow and expensive, it does allow very thick boards to be produced. These are used for ledgers, suitcases and as stiffeners in boots and shoes.

All the boards considered so far fall into the category of high substance papers. They contain one type of pulp whether that pulp be straw, waste paper, mechanical wood, chemical wood, or a mixture of these and the surface of the board has the same composition as its middle and its back. Another large group of boards are built up from layers of different composition. This lamination may be carried out on a cylinder machine, where the various vats contain different pulp mixes (page 83), but it can also be done by running together the wet webs from two or more fourdrinier wires. Using this principle, it is possible to put a layer of good quality paper on to a poor quality body. For example, a lining of a sulphite pulp with filler can be combined with a chipboard body to produce a *white lined chipboard*, with good printing quality and adequate strength for carton work. *Triplex* boards take the principle one stage further since both the back and the front of the board are lined with better quality material. Large quantities of folding box boards of various types are produced on cylinder machines. A range of *coated box boards* is available for high quality carton work.

SPECIALITY PAPERS

Many papers are produced for very specialised applications and do not fall into any of the main classes so far considered. This miscellaneous group includes crepe, waxed or cloth-lined papers and papers for a great variety of uses ranging from bibles to banknotes, and air mail stationery to fruit wrapping.

9. Paper and board – testing methods

Although a knowledge of raw materials and methods of manufacture is essential to a proper understanding of paper and board, the printer is primarily interested in the properties and the price of the finished paper. Providing these properties allow the paper to perform well on the printing press, in any conversion process and finally in use as a printed product, it does not really matter to him how the papermaker has achieved this result. In other words, although a printer may reasonably ask for a paper with a given set of properties, it would be quite unrealistic for him to tell the papermaker what this paper should contain and how it should be made.

The previous chapter outlining some of the main classes of paper and board can be read in a very short time, but learning to distinguish between these papers takes much longer and requires experience in handling the various grades. With this experience it becomes possible to say something about the composition and properties of a particular paper merely by looking at it. However, a more accurate identification and a fuller description can only be made after carrying out some tests on the paper. These tests may require no apparatus and be very rough, or they may involve scientific instruments and be extremely precise; they may be very simple or they may be quite sophisticated. Obviously the nature of the tests carried out must depend on circumstances. In carrying out fundamental research it may be necessary to describe the properties of a paper in the most precise terms and all the facilities of a modern paper testing laboratory will be used. On the other hand, an individual buying or selling paper may have to make a snap judgement on the merits of a paper without any of these facilities. Somewhere in between these two extremes of paper testing come the routine checks made on incoming materials in the control laboratory of a printing firm. In this case, although it may be possible to carry out certain selected tests usually involving simple apparatus, it may not make economic sense to include long complicated tests on expensive research instruments.

SUBJECTIVE METHODS OF TESTING

Before considering the range of instruments that have been developed for measuring the properties of paper, mention must be made of some of the ways in which information on a particular paper can be gleaned by purely subjective means using the five senses.

Quite a lot can be learned about a piece of paper simply by looking at it. If it has an ordinary machine finish it will be possible to see a wire mark on the underside. This will be more difficult to see if the paper has been well calendered as this tends to obliterate these markings. If the paper was made on a twin wire machine, the two undersides will have been brought together and there will be no visible wire mark.

Because there is a tendency for fibres to attempt to line up along the length of the moving wire, the appearance of the paper surface may show which is the machine direction. However, if this is not clear then there are a number of simple ways of finding this out. If two strips of equal size are cut from the paper at right angles to one another and held horizontally at their ends, one strip will sag more than the other. The machine direction runs along the length of the strip that sags least and across the width of the other strip. Another method of finding the machine direction is to drop a small circular disc of paper on to the surface of some water. As an increase in water content causes paper to expand more across the web than along it, the effect of wetting one side of a paper is for it to curl up into a tube running in the machine direction. A number of simple methods for determining the machine direction and the top side of paper and board are set out in the BPBMA Proposed Procedure No 57.

If a paper is held up to the light it is possible to judge whether the sheet is closely made and also to see any impurities which may be present. When two papers are being compared it is essential that they are placed side by side and not one on top of the other. Care should also be taken to ensure that the two sheets are being viewed under identical conditions of illumination and background. The opacity of two papers can be roughly compared by laying them on to a printed page and attempting to view the print through the papers.

An indication of the degree of sizing can be obtained by drawing a pen across the surface of a paper. If the writing ink feathers or penetrates through to the underside, the paper is either unsized or only soft sized.

A good deal of information can be gained by tearing a sheet of paper. The resistance to tearing will give a useful indication of strength and furnish, although it must be borne in mind that since fibres tend to line up in the machine direction, papers have better tearing strength in the cross direction than they have in the machine direction. The carefully torn edge of a paper will give an indication of fibre length and will also show whether the paper consists of several different layers, as for example in a triplex box board or whether it has the same composition all the way through.

It is sometimes difficult to decide whether a paper is coated or whether it is simply heavily loaded and uncoated. If the bevelled rim of a coin is drawn across the surface of a coated paper it will leave a distinct mark whereas it will have little effect on an uncoated paper. If coated surfaces are rubbed together white powder will sometimes come away from the sheets.

Ears as well as eyes and hands can be useful in roughly assessing the properties of a paper. If a sheet of a weak paper is shaken with both hands it will make no sound, but a strong well-sized paper will give a distinctive 'rattle' which is another indication of quality.

With experience, these simple subjective methods of assessing the properties of paper can become extremely effective and useful, particularly to those concerned with the buying and selling of paper. While these individuals usually have access to the facilities of a paper testing laboratory they are often working in situations where they have to become their own laboratory, making rough assessments based on experience. Over a period of time they can check these estimates against laboratory results and in this way their judgement can be refined. Whilst accepting the usefulness of these rather crude methods, it must be immediately said that they have severe limitations. The value of the results depends entirely on the judgement and skill of the individual and there is plenty of scope for 'operator error'. Small differences cannot be detected, numerical values cannot be given to properties and as the tests are not carried out under standard conditions, misleading results can be obtained.

OBJECTIVE METHODS OF TESTING

The obvious need for precision in measuring the properties of paper has led to the development of a wide range of paper testing instruments and methods. Most of these tests have been developed by papermakers, and in most cases one might add 'for papermakers'. Although some of these tests provide information useful to the printer, it is only comparatively recently that a serious effort has been made to develop testing methods which directly relate to the performance of a paper on a printing press. Progress in this direction has been limited, but at least the simple truth has emerged that one good method of testing the printability of paper is to print on it.

Perhaps more than any single paper property the printer needs consistency of quality, both within a delivery and between different makings. It is sometimes said that most printers can print on most papers, but although the correct press conditions for a particular paper can be established at the start of a run, variations in the properties of paper being fed through a press can lead to expensive stoppages and increased waste. Unfortunately the variations between two sheets, two reels or even different areas of the same sheet of paper can never be completely eliminated because by its very nature paper, as we know it today, is a non-uniform material. If a tin of paint or liquid ink is thoroughly stirred and then small samples are taken from different places in the tin, the samples will be found to be largely identical. The same would be true for samples taken from different tins of paint or ink providing that they all came from the same making. These are uniform materials and their composition does not vary appreciably within a single making. This is not the case for paper. Small samples taken from different parts of a sheet do not have the same composition and no two sheets of paper are ever identical. The printer will have to go on accepting these variations for as long as paper is made from natural fibres and by existing methods of manufacture. The future holds the possibility of far more uniform papers made from synthetic polymers by quite different methods.

SAMPLING FOR PAPER TESTING

Because different sheets of the same making of a paper are not identical, tests carried out on a single sheet are clearly of limited value. Clearly it would be absurd to go to the other extreme and test every single sheet in a delivery. Apart from the time and cost of such an operation, the destructive nature of some of the tests would leave no paper to print. The only practical solution is to carry out tests on a sample of the paper, to obtain an average value of the results from several sheets and an indication as to whether the measured value for any one sheet falls outside acceptable limits for a particular paper. For example, in measuring the thickness of a board ordered at 500μm, it might be found that although the average thickness of the sheets sampled was acceptable at 515μm, the sheets taken from a particular pallet fell outside acceptable limits at 550μm.

The actual sampling method adopted must depend on circumstances. There will be occasions when only a small sample is available for testing. Truly representative sampling of paper involves the selection of a certain minimum number of reams, pallets, etc, from a consignment, the withdrawal of a certain minimum number of sheets from each of these units, and the selection and cutting of specimens, from which will be taken the samples necessary for the various tests. Details of the British Standard method for the sampling of paper and board for testing can be found in BS 3430.1968. When taking sheets from a stack or ream, allowance should be made for the fact that since sheet cutting machines are often fed by several reels simultaneously, similar sheets may recur at fixed intervals in the stack (page 89). Some

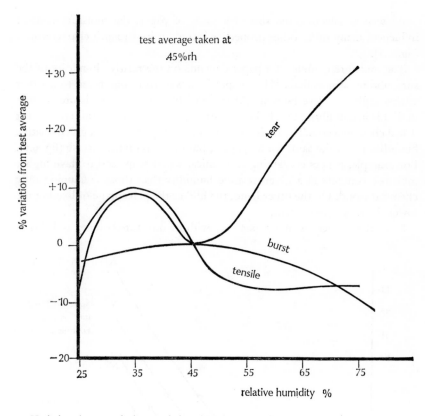

test average taken at
45%rh

9.1 Variations in strength characteristics with relative humidity (PIRA).

knowledge of statistical method is essential to an understanding of the importance of sampling and to a proper interpretation of test results, and the notes set out in BS 2987.1958 provide a useful introduction to the application of statistics to paper testing.

HUMIDITY AND PAPER TESTING

We have seen in earlier chapters that the size of a sheet of paper changes with its moisture content. Cellulose fibres have an affinity for water and on taking it up they swell causing the paper to expand in all three dimensions. The increase in the diameter of the fibres is proportionally much greater than their increase in length, so a paper expands much more in thickness and in the cross grain direction than it does in the grain or machine direction (page 91).

As well as affecting the size of a piece of paper, the moisture content influences many of its other properties including its strength characteristics (fig. 9.1).

The moisture content of a paper depends on the relative humidity of the surrounding atmosphere. When paper is moved from one room to another with a higher relative humidity, it will slowly increase its moisture content until it is in equilibrium with the new atmosphere. In making this adjustment it will change in size. The moisture content corresponding to a given relative humidity is not the same for all papers, but depends on the furnish (fig. 9.2). For example, papers containing mechanical wood pulp tend to have higher moisture contents at a given relative humidity than those containing only chemical wood. On the other hand, the higher the percentage of loading the lower the moisture content.

Since the properties of a paper vary with its moisture content and this in

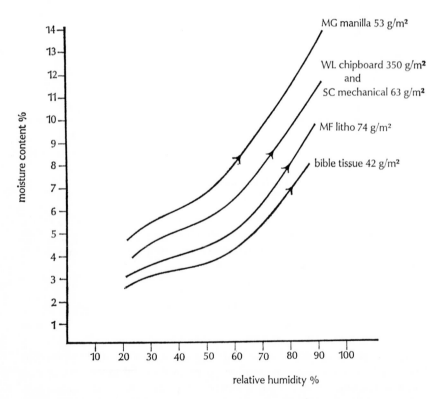

9.2 Relationship between moisture content and relative humidity for various papers.

turn depends on the relative humidity of the surrounding atmosphere it follows that paper tests should be carried out under standard atmospheric conditions. The current British Standard and Primary ISO 'standard' conditions for paper and board testing are 20°C/65% rh. In the United States and some other parts of the world the standard preferred is 23°/50% rh. Arguments for this lower relative humidity include the facts that rh changes have less effect on paper properties at 50% rh than they do at 65% rh, that the increasing use of central heating in offices, factories and shops makes 50% rh more typical of end use conditions than 65% rh, and that heated un-conditioned laboratories used for paper testing often have relative humidities in the range 40–60%.

Although it may not always be possible, it is desirable that before paper is tested, it is stored for a period of several hours under standard conditions so that there is time for its moisture content to come into equilibrium with these conditions.

PAPER TESTING METHODS

A great many methods of testing the properties of paper have been developed over the years and some of the more important of these are now outlined. They are not dealt with in order of their importance to the printer, but the emphasis has been placed on those properties which influence the printing performance of a paper. A section of the appendix provides references for those seeking more detailed information on some of the testing methods and also indicates where the various items of equipment referred to in the text may be obtained.

Grammage (basis weight or substance)

The grammage or substance of a paper is the weight of a given area of that paper. The units for substance are grammes per square metre (g/m^2), although traditionally it has been expressed as pounds per ream of a stated sheet size. Paper substance is found by weighing a sheet of paper of known size. This is normally carried out on a quadrant balance (plate 7). A template is used to prepare a small standard square of paper, which is then placed in a cradle suspended on one arm of the balance. The substance of the paper in grammes per square metre is indicated by the other arm moving across the quadrant scale.

In the laboratory, substance may be measured more precisely by weighing a sheet or sheets of paper of standard size, *eg* 100cm × 100cm on an accurate balance. Standard methods for measuring the substance or grammage of a paper are set out in BS 3432: 1971.

Since the printer normally buys his paper by weight, but is then primarily interested in the area which he has to print, substance is an essential measure.

Thickness (or caliper)

The thickness of paper is measured by a micrometer and the result expressed in micrometres (microns) μm (0·001mm). Since paper is a compressible material, it is important to exert a steady uniform pressure on the samples and hence deadweight dial micrometers (plate 8) are preferred to spring loaded or ordinary hand micrometers, which can give misleading results. The standard method set out in BS 3983:1966 specifies a deadweight dial micrometer exerting a pressure of 49·3kPa. A pack of ten sheets is placed in the jaws of the micrometer and measurements made at five positions on the wad. The average value of the five readings is then divided by ten to give the thickness of the sheet. The foot of the micrometer should always be lowered gently on to the sample in order to avoid compressing the paper.

Serious variations in the thickness of paper during a printing run will lead to differences in impression on the printing machine and hence to density variation between prints. Problems can also be caused in later processes, for example on carton making and carton filling machines.

The *bulk* of a paper, the inverse of its density, relates the thickness of a paper to its substance and gives an indication of the ratio of air space to solid matter within a sheet.

$$\text{Bulk} = \frac{\text{Thickness in micrometres, } \mu\text{m}}{\text{Substance in grammes per square metre, g/m}^2}$$

Strength

Whilst substance and thickness are both fundamental properties which can be precisely defined in terms of the dimensions and mass of a paper, most paper testing methods have been developed empirically and the value given for a particular property will depend on the method used and the conditions which apply. For example, a number of instruments have been developed to measure the smoothness of a paper surface, but because of differences in their design they are not all measuring precisely the same property. In the same way, there is no single measure of the strength of a paper, but tests have been developed to provide information on different types of strength, the more important of these being tensile, bursting, tearing and folding strength.

Generally speaking these tests are of more value to the papermaker than to the printer, but there may be circumstances when the demands made on the

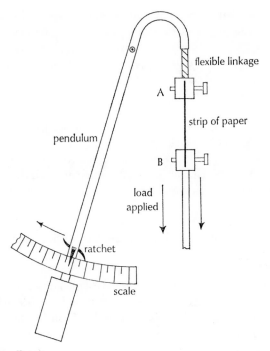

flexible linkage

A

strip of paper

pendulum

B

load
applied

ratchet

scale

9.3 Principle of tensile tester.

paper during the printing process or in use as a printed product make it essential that a particular strength characteristic does not fall below a certain minimum value. For example, one can see the importance of the tensile strength of a paper which is going to be subjected to the tensions of a web-fed press, the bursting strength of a paper bag or sack, the tearing strength of a playing card or the folding strength of a currency paper. Furthermore, taken together these testing methods can give a useful indication of the overall quality of a paper. In making any of these strength measurements, care must be taken to avoid samples which are already weakened by creases, holes or other imperfections.

Tensile strength

The tensile strength of a paper is a measure of the load which has to be applied to the ends of a strip of paper for it to break. The strip of paper, normally 15mm wide, is fastened between two clamps A and B (fig. 9.3). The load applied in the direction of the arrow at B, pulls the weighted pendulum along the scale until the paper breaks. At this point a ratchet holds the pendulum so that its position can be read on the scale. The Schopper tensile tester (plate 9) operates on this principle and applies a gradually increasing load to the test sample. The tensile strength is expressed in N/15mm width. Static tensile testers of this type are normally also equipped to measure the amount of stretch in the paper strip at the moment of break. The determination of the tensile strength of paper and board is the subject of BS 4415:1969.

It has been shown that the actual working performance of paper made into bags or sacks is related more closely to its dynamic tensile strength, which is a measure of its ability to stand up to sudden stress. This property may be determined on Van der Korput or Bekk instruments which apply a sudden load to the test strip and measure the work done in breaking it.

Bursting strength

If a rubber diaphragm, held firmly in contact with a sample of paper, is gradually inflated, the paper will be forced out until it eventually bursts. The pressure developed behind the diaphragm at this point is a measure of the bursting strength of the paper (fig. 9.4). Standard methods based on this principle are set out in BS 3137:1959, the pressure behind the diaphragm either being developed hydraulically or pneumatically. In the Mullen Burst Tester (plate 11), the paper sample is firmly clamped by a ring 30·48mm in diameter and hydraulic pressure is developed behind the diaphragm by steadily turning a handle which drives a piston into the liquid (glycerine or ethylene glycol) which is in contact with the diaphragm. A gauge records the pressure at the moment of burst, and the bursting strength is expressed as the mean of 20 readings in Kg/cm².

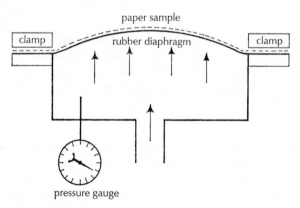

9.4 Principle of burst tester.

Tearing strength

We have already referred to the fact that useful information can be gained simply by tearing a piece of paper (page 108). A numerical value for the tearing strength of a paper can be found using the Elmendorf Tear Tester. Test samples of standard dimensions are first prepared on a special guillotine. These samples are small rectangular pieces of paper with two cuts made into one of the long edges (fig. 9.5). The instrument itself consists of a heavy sector-shaped plate, which can swing freely from its apex (plate 10). When held

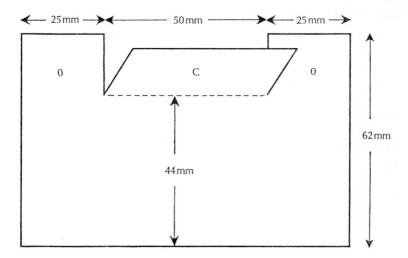

9.5 Test sample for Elmendorf tear tester.

with one of its edges in a vertical position, the central tongue C of the test sample can be held in a clamp attached to the raised edge of the plate and the outer tongues O can be held in the fixed clamp above the plate. When the plate is released from this raised position, the tearing paper sample acts as a brake on its pendulum swing and prevents it from reaching the same height on the other side. The extent to which the swinging plate falls short of this height is indicated by a pointer on a scale round the circumference. This distance is directly related to the braking effect of the sample and is a measure of the tearing strength of the paper. Tests must be carried out on samples cut both in the machine direction and the cross direction. Ten readings should be taken in each direction. Two or more sheets may be torn at any time, and the tearing resistance may be expressed as the mean value multiplied by a pendulum factor (commonly 3) to convert it to mN, divided by the number of sheets torn at a time. Standard methods of determining the tearing resistance of paper are described in BS 4468:1969.

It is important to realise that the Elmendorf tester does not measure the initial tearing strength of a straight edge of paper, but the resistance to tearing of a tear already started. The edge tearing strength of paper can be measured on the Concora torsion tear tester.

Folding resistance

If a strip of paper is subjected to continuous folding under tension it will eventually break. The number of folds which cause this break provides a measure of the folding resistance of a paper. This principle is used in the Schopper instrument which is most commonly used to measure this property. A standard test procedure for the measurement of folding resistance is set out in BS 4419: 1969. As with tensile and tearing strength, tests must be carried out in both the machine and the cross direction of the paper.

Optical properties

The appearance of a paper depends on its optical properties including its opacity, its brightness, its colour and its gloss. Each of these properties can be judged by an experienced eye, but the need for more precise measurements has led to the development of testing methods, many of which involve the use of optical instruments.

Opacity

The opacity or absorbance of a sheet material can be regarded as its light stopping power, the extent to which the sheet stops the passage of incident light. However, in considering the properties of printing papers, we are more interested in the extent to which the sheet will hide what is beneath it. For example, we want to avoid envelope papers which allow one to read the contents of an enclosed letter, and book papers which allow pages 1, 2 and 3 to be read simultaneously. This hiding power of a paper is known as its printing opacity. It differs from its optical opacity or absorbance because a proportion of the light passing through the paper is scattered (fig. 9.6). The effect of this scattering is that rays of light coming from an object underneath the paper are deflected from their course and fail to reach the eye. The greater the amount of scattering, the more the image will be obscured. The more reflecting surfaces available in the paper, the higher will be its printing

118

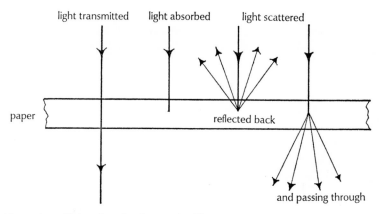

light transmitted   light absorbed   light scattered

paper

reflected back

and passing through

9.6 Scattering of light reflected and transmitted by paper.

opacity. These reflecting surfaces are provided by paper fibres and by particles of filler like china clay, so printing opacity is favoured by heavily loading a paper or by preserving an open fibrous structure. On the other hand, heavy beating, supercalendering or wax coating have the reverse effect, since they each tend to reduce the number of reflecting surfaces available. Printing opacity is defined as the ratio of the amount of light reflected from a single sheet of paper with a black backing to the amount reflected from a pile of sheets of the paper, thick enough to be opaque.

One simple way of roughly assessing the opacity of a paper is to find the number of sheets that have to be laid on top of a standard mark or print on another paper, before the mark is no longer visible. Another simple method developed by PIRA some years ago is based on a numbered set of seventeen standard papers heavily printed with black ink, each having measured degrees of printing opacity from 71 to 100 per cent. The paper under test is placed over a portion of the print on the paper with 100% opacity, and a paper is selected from the standard range which obscures the print to the same extent. This method is completely subjective and its accuracy limited by the opacity intervals in the range of standard papers. However, it does provide a useful means of assessing the amount of show-through associated with a certain printing opacity.

Instruments with varying degrees of sophistication are available for the more accurate measurement of printing opacity. The Eel Opacimeter (fig. 9.7) is a simple and reasonably inexpensive instrument which is widely used in paper and printing control laboratories. Light from a normal tungsten lamp provides even illumination for the underside of the sample on the platform at

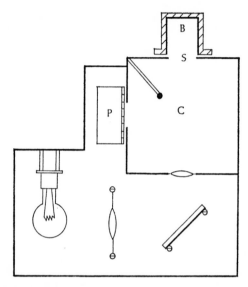

9.7 Principle of the EEL Opacimeter.

S. Light reflected back from the sample is scattered within the white lined cube C and then falls on the photocell at P. The resulting current from this photocell is measured by a galvanometer. The instrument is first adjusted so that a thick wad of paper under test gives a scale reading of 100. The wad is then replaced by a single sheet of paper, held down by a black lined cap B, when the reading of the galvanometer is a measure of the percentage printing opacity.

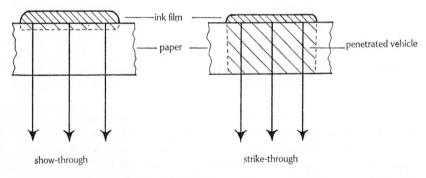

9.8 Show-through and strike-through.

Paper and board – testing methods

When the printing on one side of a paper can be seen through the paper, this print-through may either be due to *show-through* or *strike-through* (fig. 9.8). Show-through is related to the printing opacity of the paper but strike-through is the result of the vehicle of the printing ink penetrating right through the sheet. When paper absorbs an oily or resinous substance it becomes more transparent, so strike-through makes it easier to see a print on the reverse side of the paper. Normally the vehicle only penetrates a short distance into the sheet, but even this will have the effect of reducing the printing opacity, so making print-through more likely.

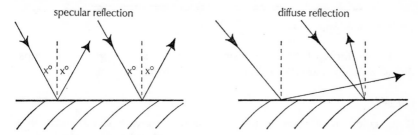

9.9 Specular and diffuse reflection.

Gloss

As every schoolboy knows, rays of light striking the surface of a mirror are reflected in a regular way, so that the angle of incidence is equal to the angle of reflection (fig. 9.9a). This is called *specular reflection*. The paper, on which these words are printed, also reflects light but because of the unevenness of the surface the light is not reflected in a regular way but is scattered in all directions (fig. 9.9b). This is called *diffuse reflection*. We are able to see most everyday objects by the light which is diffusely reflected from their surface. Although paper can never be given a mirror-like finish, processes like cast-coating (page 98) and friction glazing (page 88) can give a highly polished surface, which will specularly reflect a good proportion of the incident light.

The gloss of a paper depends on the amount of specular reflection from its surface and it can be measured by comparing this amount of light with that reflected from a standard glossy surface. The principle of a simple glossmeter is shown in fig. 9.10. The instrument is first placed on a polished black tile, and the current from the photo-electric cell causes a deflection of the galvanometer, which is adjusted to read 100 units. The tile is then replaced by the paper sample, and the reading of the galvanometer is the specular reflection or gloss of the paper, expressed as a percentage of that of the standard tile.

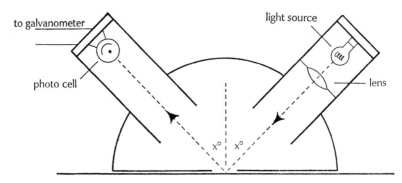

9.10 Principle of simple glossmeter.

As the two limbs of the instrument are set at the same angle to the sample, diffuse reflections are not collected by the photocell.

Colour

If the colour of a printing paper is allowed to vary significantly, these variations can easily show themselves in the printed product, for example when the separate sections of a book are of noticeably different shades of white. To avoid the bad impression created by these differences, the paper-maker must ensure that the colour of a particular paper is held within certain acceptable limits.

In practice most colour measurement is done with that remarkable instrument called the eye. Fortunately, in making paper and in printing, all that is normally required is to be able to judge whether two samples are properly matched. An experienced colour matcher can do this remarkably well, normally at least as well as the eventual customer and user. The two samples being compared must be viewed side by side under standard conditions of illumination. When experience is combined with reasonable care, this subjective method of assessing colour is adequate for many purposes, but of course it does have serious limitations. To begin with, no three eyes are the same, even if two sometimes are, and so it may be that a papermaker or printer and his customer do not see 'eye to eye' over the colour of a paper or print. Defects in colour vision are common, and colour fatigue can cause some odd effects. These limitations of visual colour matching have led to the development of instruments which will 'look at a colour' and describe it in figures or in the form of a graph.

It should be pointed out that in defining the colour of a surface we are really concerned with three different factors, hue, saturation and brightness.

Hue may be described as that aspect of visual perception that allows us to distinguish red, green, yellow, etc, and is given by the dominant wavelength of the monochromatic light. Saturation depends on the amount of white diluting a colour. Light of a single wavelength is said to be fully saturated. Brightness is the ratio of the amount of light reflected from the specimen compared with that reflected under the same conditions from magnesium oxide.

A detailed study of the physics of colour and methods of measuring colour is beyond the scope of this book, but fuller information can be found in the references listed in Appendix A. Colour measuring instruments are all based on the general principle of illuminating the sample with coloured light of a particular wavelength and measuring the amount of that light reflected from the surface. In a *continuous recording spectrophotometer* the colour of the light falling on the sample is changed continuously across the spectrum, and the instrument traces out a graph indicating the extent to which light of various wavelengths is reflected from the sample (fig. 9.11). These continuous recording instruments are very expensive and are essentially for use in a research laboratory. The Zeiss Electric Reflectance Photometer (Elrepho) is a manually operated but extremely sensitive instrument, named in BS 4432:1969 as being suitable for carrying out standard methods of determining the optical properties of pulp, paper and board. The basic construction of the instrument

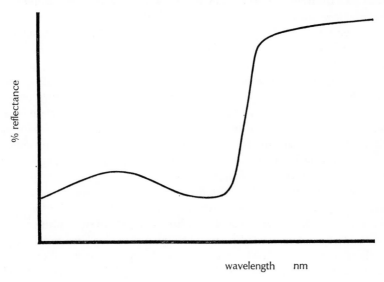

wavelength   nm

9.11 Spectrophotometric curve for a red print.

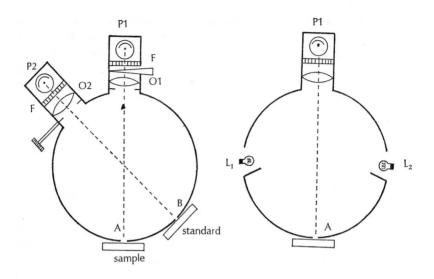

integrated sphere painted matt white

9.12 Principle of the Elrepho reflectance photometer.

can be seen from the two sectional views shown in fig. 9.12. The sample to be measured is applied to a 35mm opening, A, in the photometer sphere, and is illuminated by diffused light originating from lamps L_1 and L_2. A standard white plate is applied to the second opening at B. An objective O1 and photo-cell P1, placed directly opposite A, only receive light reflected from the sample at A, and in the same way an objective O2 and photocell P2 only receive light reflected from the standard plate at B. The filters F positioned in front of the photocells can be simultaneously changed by a revolver mechanism, and the amount of light reflected from the sample is compared with that reflected from the standard, for any given pair of filters. One of the advantages of the instrument is that the sample is diffusely illuminated, and so is being 'seen' in the way that materials normally are seen.

This is not the case in the Eel reflectance comparator, a simpler and less expensive instrument which is widely used for control purposes in the printing and paper industries. Fig. 9.13 shows how light from a small filament lamp is concentrated by a lens and passed through a coloured filter to fall on the sample at an angle of 45°. The filter only lets through a portion of the visible spectrum. If any of this coloured light is reflected by the print, some of it will fall on the photo-electric cell placed directly over the print. The photocell is put in this position, rather than at 45°, so that any specular reflection from the print due to its gloss is not collected. The cell converts the light energy received into a small electrical current which is recorded by a micro-ammeter.

In use, the comparator head is placed first on to a standard white surface (normally a block of magnesium carbonate) which is assumed to reflect 100% of the visible light of all wavelengths. The micro-ammeter is adjusted so that the scale reading is 100 units, then the comparator head is moved on to the surface being measured. If the reading is only 25 units, this indicates that only one quarter as much of the light passing through the filter is reflected by the sample as was reflected by the standard white surface. The colour filter is now replaced by a second filter which lets through another portion of the spectrum. The comparator is again standardised on the white block and a new reading obtained from the sample. This method is repeated with a range of filters which cover the whole of the visible spectrum. Knowing the band of wavelengths that each filter transmits, it is then possible to draw a graph of percentage reflection against wavelength. The graphs obtained look similar to those produced by a continuous recording spectrophotometer but the information they contain is more limited. A number of different sets of filters are available for the comparator head, and the instrument may be used to measure brightness as well as colour.

Moisture content

Dry paper is an extremely hygroscopic material, that is to say, it has a strong tendency to absorb moisture from the atmosphere. The amount of moisture it absorbs depends on the amount of moisture vapour held in the surrounding

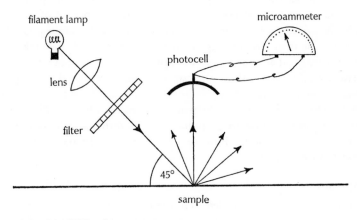

9.13 Principle of the EEL reflectance comparator.

air, given by its relative humidity. It also depends on the furnish of the paper and on beating and other conditions in its manufacture. In other words, different papers kept under the same atmospheric conditions are likely to contain different amounts of moisture. This is well illustrated in Table 9.1

Table 9.1   % Moisture content at varying relative humidities

Relative humidity	43%	65%	82%
Type of paper			
Sulphate sack	7·46	9·44	12·3
Woodfree	8·00	9·75	12·7
Glassine	8·76	10·5	13·5

which shows the moisture content of several papers at three different relative humidities and also by fig. 9.2 on page 112.

A paper's moisture content influences all its other properties, some more than others. From a printer's point of view, its effect on the dimensions of the paper cause the most problems, but variations in moisture content can also lead to other difficulties in printing and finishing, eg retarded drying of ink, variations in the creasing properties of carton board.

The moisture content of a paper or board is normally measured by weighing a sample on an accurate laboratory balance and then drying it in a ventilated oven at 102–105°C until it reaches a constant weight. The moisture content is expressed as the percentage loss in the original weight of the sample. Although this is a simple method, it is one that demands careful sampling and accurate working if the result is to have any real value. Details of a standard procedure are set out in BS 3433:1961. The weighing should preferably be carried out in an airtight bottle or tin, and after the period of drying, the sample should be cooled in its closed container in a dry atmosphere before weighing. The main disadvantage of the method is that it may take several hours. Where great accuracy is not essential a number of more rapid methods are available. In one of these the paper is simply weighed on a quadrant balance before and after drying. In another, the sample is dried by an infra-red bulb which is fixed above the pan of an enclosed balance (fig. 9.14). As the sample gives up moisture, a pointer moves across a scale indicating the percentage loss in weight. Moisture content can also be measured using an electrical moisture meter.

Temperature

The amount of moisture needed to saturate a given volume of air increases as the temperature rises. If warm air containing a certain amount of moisture

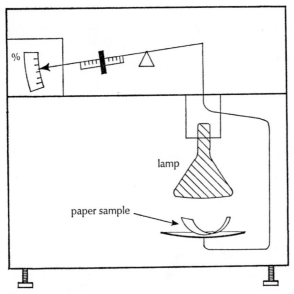

9.14 Moisture content balance.

is cooled, its relative humidity rises and the dew point may be reached, when the moisture cannot all be held as a vapour and water condenses out on to any available surface. A car window mists up because some of the warm moist air inside the car is cooled by the cold glass to the point when water condenses out in small droplets on the inside of the window. In a similar way, if a cold stack of paper from a warehouse is moved into a warm machine room, the air close to the stack may be cooled below its dewpoint so that water condenses out on to the edges of the sheets, causing expansion, wavy edges and consequent troubles on the printing machine. To avoid this, a cold stack of paper should not be unwrapped until its temperature has reached that of the machine room. This can be checked by measuring the temperature of the stack with a PIRA stack thermometer (plate 12). The long stem of the thermometer should be inserted as low down the stack as possible through a small slit made in the outer wrapper, and left for 15 minutes. If there is a significant difference between the stack and air temperatures then the slit in the wrapper should be sealed, and the stack given more time to reach the machine room temperature.

Dimensional stability

In colour printing, one of the printer's greatest difficulties is keeping paper

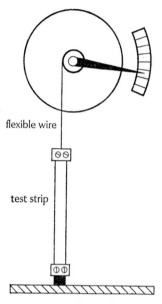

flexible wire

test strip

9.15 Simple instrument for testing the dimensional stability of paper.

the same size during successive printings. Although a sheet of paper may be subject to various mechanical stresses during the processes of printing and finishing, changes in its size are usually due to variations in its moisture content. These variations in moisture content correspond to changes in the relative humidity of the air surrounding the paper. We have already seen that the extent to which a sheet of paper changes its size with changing humidity is not the same for all papers (page 112). One simple way of making a rough comparison between the dimensional stability of one paper and another is to measure the increase in length when equal strips of the two papers are totally wetted. Fig. 9.15 shows a simple instrument which can be used to make these measurements. A test strip cut across the machine direction is held between a fixed clamp and a second clamp connected to a pulley and pointer, arranged so that any increase in the length of the paper is greatly magnified in the movement of the pointer across the scale. Scale readings taken when strips of different papers are totally wetted provide a rough comparison of the dimensional stability of the papers. More sophisticated instruments are available in which test strips are clamped in glass containers through which air of chosen relative humidities can be passed and changes in length of the paper can be accurately measured by micrometers.

These methods of measuring dimensional stability can be useful, but normally the printer simply wants to know whether the paper he is about to print is in moisture balance with the machine room conditions. This can be checked with the aid of the PIRA Paper Equilibrium Tester (PET), a sword-shaped duralumin bar incorporating two needles A and B 50·8mm

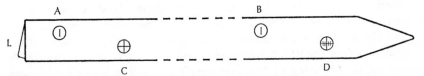

9.16 PIRA paper equilibrium tester (PET).

(20 in.) apart and two glass graticules C and D also 50·8mm apart (fig. 9.16). The sword is thrust through the wrapping into a ream or stack of paper. Pressure on the lever L at the end of the bar raises the needles at A and B so that they make two reference marks on the sheet of paper. Inked felt pads around the needles make it easier to see the two reference marks. This marked sheet of paper is taken from the ream and hung in the machine room for 15 minutes, so that it is able to come into moisture balance with the surrounding atmosphere. The sheet is then laid on a flat surface, the two reference marks are viewed through the two glass graticules and the graduated scale on the window D indicates the change in the original distance between marks AB. With experience of the test, it is possible to lay down limits on the degree of expansion which will cause register problems on a particular type of job. Where these limits are exceeded the paper should be conditioned before being printed. As a rough guide, an expansion of 0·5% or more is likely to lead to wavy edges if the paper is left unwrapped in the machine room for any length of time.

Surface strength

On a letterpress or litho machine only about half of the ink carried on the relief surface or offset blanket is actually transferred to the paper at each impression. In other words, the ink film is split, part of it being transferred and part of it remaining on the printing surface. The tack of the ink is a force resisting this film splitting, and this tack may be so strong that splitting takes place at the weakest point in the paper instead of in the ink film (fig. 14.7). Since the tack of a printing ink increases with the speed of printing, paper surfaces are subject to greater stresses on fast-running machines. Weaknesses in the surface structure of paper may show themselves in a number of ways, but the problems created are certainly greatest in multi-colour offset printing.

Fluffing

The pulling away of individual fibres from uncoated papers is known as fluffing. This loose material may adhere to the printing surface, so causing blemishes on successive prints or it may work its way back to mix with the ink and cause a filling-in of halftone dots.

Perhaps the best method of measuring the fluffing tendency of an offset paper is to run a printing test on a small press under standard conditions and with the aid of a microscope count the number of fibres adhering to the rubber blanket. The PIRA fluff tester provides a shorter and more practicable method, not involving a press. Five sample sheets of paper are fed between two rollers, rotating in contact under a known pressure. One of the rollers is covered with a rubber blanket, and this collects particles of fluff from the side of the paper being tested. The number of fibres on the rubber blanket is then counted and the fluffing tendency of the surface expressed in the number of fibres per square cm. Since the number of loose fibres on the underside of a paper is normally much greater than on the topside it is important to know which of these surfaces is being examined. The use of the PIRA fluff tester is described in the proposed procedure No 8 of the BPBMA.

Picking

The lifting of areas of the paper surface during printing is known as picking. The term is normally associated with coated papers, when poor adhesion of the coating to the body may result in pieces of the coating coming away. The break may also occur in the body of a coated or uncoated paper or between the laminates of a board, when the term *splitting* is sometimes used.

Unfortunately the picking of material from a single sheet of paper can lead to a large amount of printed waste, because of the effect on following sheets. The fragment of paper may stick to the printing plate or blanket, and after a few impressions may itself start to print. Since it is standing higher than the general level of the printing surface the small area printed is surrounded by a white unprinted ring or halo. These characteristic marks on printed sheets are known as secondary picks or 'hickies'.

The most reliable method of measuring the surface strength of a paper uses the IGT Printability Tester. This versatile instrument was developed by IGT, the Dutch Printing Research Institute, in an attempt to reproduce, as closely as possible, the actual conditions operating on a full scale printing machine. Its versatility lies in the fact that it allows a study of the printing performance of a paper at different printing speeds, at different printing pressures and with different inks. Apart from its value in measuring pick resistance it can also be used to investigate smoothness, absorbency, ink drying rate and many other printability variables. The IGT apparatus consists of two basic units, one being a simple roller system for distributing a measured amount of ink and for inking up a small metal disc, and the other a printing unit on which ink is transferred from this disc to a small strip of paper.

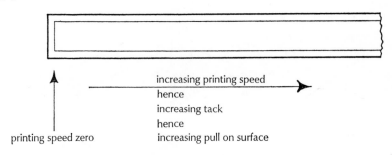

printing speed zero

increasing printing speed
hence
increasing tack
hence
increasing pull on surface

9.17 IGT printability test strip.

In measuring pick resistance, 1cm³ of a special oil of known viscosity and tack is transferred to the distribution unit with the aid of an ink pipette. The unit (plate 13) which comprises two steel and one polyurethane roller, must be run for at least eight minutes, during which time the top roller should be turned end to end every two minutes to ensure even distribution. The printing disc is then run in contact with the top roller for 45 seconds, before it is ready to be transferred to the printing unit (plate 14).

Strips of paper, 25cm long and 3cm wide, are cut out in both grain and cross grain directions and carefully marked. One of these strips is clamped round the 150° impression sector, which is the counterpart to the impression cylinder of a printing machine. The printing pressure is adjusted by means of a handle at one end of the printing unit. The heavy pendulum, which is connected to the impression sector, is then released and falls at an increasing speed. Since the tack of the oil increases with the increasing speed of printing, so does the pull exerted on the surface of the paper strip. A point may be reached when this pull becomes too strong for the paper surface and picking commences. A binocular microscope with low angle illumination may be used to judge more accurately the first point on the printed strip where the paper surface has been lifted. The distance from the beginning of the print to this point can be used to calculate the printing speed at which picking commences and so provides a measure of the pick resistance of the surface (fig. 9.17). PIRA recommends that, paper for letterpress printing should show no picking before 70cm/s and for offset processes no picking before 105cm/s. The

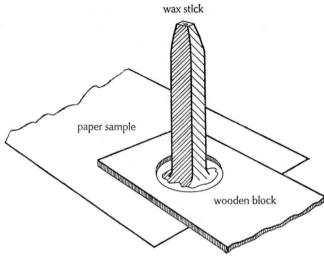

wax stick

paper sample

wooden block

9.18 Dennison wax test.

measurement of surface strength using the IGT printability tester is described in the BPBMA Proposed procedure No 20.

The main drawback with this IGT method is that it involves the use of an expensive instrument. An alternative, but much less satisfactory method is provided by the Dennison wax test. A set of eighteen sticks of blended waxes provide a range of varying grades of 'adhesiveness'. Each stick is numbered and coloured to allow easy identification of a particular grade. In making a test, the ends of a number of suitable wax sticks are melted in a flame and pressed on to the surface of the paper while still molten. After leaving the sticks to cool and solidify for about 15 minutes, they are pulled away from the surface, the paper being held down by a wooden block with a hole in it (fig. 9.18). The flattened base of each stick is then carefully examined for any sign of paper adhering to it. The higher the number of the wax, the greater the pull exerted on the paper. The number of the weakest wax to break the surface represents the pick number for that paper.

The Dennison method bears no resemblance to the printing process so it is hardly surprising to find that results do not offer a completely reliable guide to the picking resistance of paper on a press. Some misleading results can probably be accounted for by the effect of the high temperature of the molten wax on constituents in the paper, including resins which may be melted and moisture which is driven out. Despite its many shortcomings, the test does have the important advantages of being simple and inexpensive to carry out, and it provides a useful rough guide to surface strength. It is generally accepted that coated papers and boards for litho printing should have a pick number of at least 6. The figure for a letterpress paper can be lower than this, but the limit set must depend on the type of machine, the properties of the ink, and other press variables. Before arriving at a specification for pick resistance or for any other paper property, it is sensible to first examine the relationship, if any, between test results and press performance over a considerable period of time. The measurement of picking strength by the Dennison wax method is the subject of the BPBMA Proposed procedure No 12.

Smoothness

The surface smoothness of a paper is a major factor influencing print quality. On it may depend the fineness of the halftone screen that can be used in making the printing plate, and the amount of ink and printing pressure required on the press. If the 'valleys' of a paper surface are not properly printed because the ink fails to 'bottom', the print will have a grainy appearance. Overcoming this effect by applying more ink can result in the filling-in of the halftone dots in the darker tones. In gravure printing, research has shown that the defect known as speckle, when certain cells fail to print, is related to surface roughness. In view of the importance of this property of smoothness it is not surprising that a variety of methods have been developed to measure it, based on several different principles.

Some assessment of smoothness can be made by simply examining the surface of a paper under a low-powered binocular microscope. The relief effect can be improved by illuminating the paper at a very low angle. This technique of examination will also reveal surface defects in a paper, *eg* loose coating or fibres, coating pinholes or streaks, and it can be very useful in the investigation of print faults.

The microcontour test is widely used in the routine control of paper quality because it is simple, rapid and requires no expensive equipment. A thick film of a heavily pigmented ink of the copperplate type is rolled on to the surface of the paper and immediately wiped off. Although the bulk of the ink is readily removed, those parts filling surface depressions are held on the paper, so that the film of ink remaining provides a visual picture of the surface contours and shows up any local defects. The microcontour print may be viewed under a binocular microscope or if a numerical value for smoothness is required, reflectance measurements may be carried out (page 124).

The Bendtsen smoothness tester measures the rate at which air, under a pressure slightly greater than atmospheric, escapes underneath a thin metal

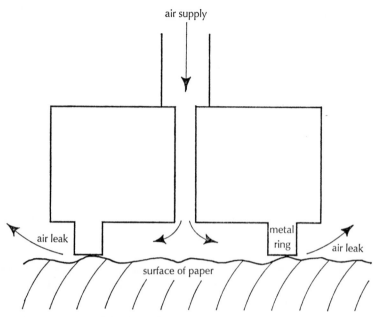

9.19 Principle of the Bendtsen smoothness tester.

ring placed on the surface of the paper (fig. 9.19). The smoother the paper, the better is the contact between the metal ring and the paper surface and the fewer gaps allowing the air to leak. The rate at which the air escapes is indicated on a manometer tube connected to the air supply. Recommendations for the determination of the Bendtsen roughness of paper and board are made in BS 4420:1969.

Although the Bendtsen test can be carried out rapidly, it suffers from the disadvantage that it does not distinguish between a paper surface with a large number of small depressions and one with a small number of large depressions, although these differences would have a marked effect on printability. Furthermore, the test results may be strongly influenced by paper properties other than smoothness *eg* air permeability and the pressure applied to the paper sample is far below normal printing pressures. In order to overcome these deficiencies, the Print-Surf roughness tester (plate 13), has been developed by J R Parker. This is an air leak instrument which measures the roughness of paper under printing pressures with a paper sample, which is backed by an actual press packing or a standard equivalent. The design of the sensing head (fig. 9.20 a, b and c) minimises air flow through the paper and protects the

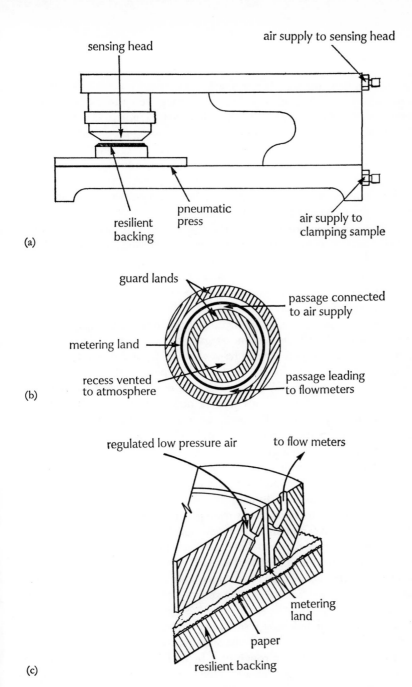

(a)

sensing head

air supply to sensing head

resilient
backing

pneumatic
press

air supply to
clamping sample

(b)

guard lands

passage connected
to air supply

metering land

recess vented
to atmosphere

passage leading
to flowmeters

(c)

regulated low pressure air

to flow meters

metering
land

paper

resilient backing

9.20 Principle of the Print-Surf roughness tester: (a) side view; (b) sensing head seen from below; (c) section of sensing head.

metering land from damage. It is claimed that the instrument can accurately predict minor changes in print quality due to roughness variations, and it is in regular use for the quality control of both coated and uncoated papers.

The Bendtsen tester and some other instruments using the air leak principle (*eg* Bekk, Gurley) can be used in a modified form to measure *porosity*, that is, the the rate of passage of air through a paper under the influence of a pressure difference. This property is of value to the papermaker since it provides information on structural characteristics. It is particularly useful when a paper is going to be coated or impregnated.

The Chapman smoothness tester is based on an optical principle. The paper sample is placed in contact with a glass prism illuminated from above. Two photo-electric cells compare the amount of light reflected from the paper as a whole with that reflected from those parts of the paper surface which are in contact with the glass. The smoother the paper, the greater the areas of contact and the closer together the two values will be. In its modified form the Chapman tester gives a visual picture of the areas of contact and non-contact, so that the distribution of the irregularities can be seen. Although the method has proved very effective in research, the Chapman method is not yet available in a form suitable for routine control purposes.

Since paper is normally being compressed at the moment of ink transfer, information on the smoothness of a paper is incomplete without a knowledge of its *softness* or *compressibility*. The Bendtsen or Print-Surf testers can be used to measure this property by comparing the rate of air leak for different pressures between the sensing head and the paper. Differences in these rates would be greatest for soft papers. In a similar way with the Chapman instrument, the pressure between the paper and the glass block can be varied and readings taken to assess the softness of the paper.

Another approach to smoothness testing is made in the Talysurf instrument, which is more commonly used to examine the surface of metals. A fine stylus is drawn over the paper and its movements amplified and traced out on a chart to give a magnified profile of a small part of the surface. The instrument is most effective when used to compare very smooth papers, but its high sensitivity, the limited distance of trace and its high cost are drawbacks to its more general use.

Other methods of measuring smoothness have been developed which depend on the fact that a given volume of liquid will spread over a larger area on a smooth paper than on a rough one. In a PIRA method based on this principle, a slot containing a fixed volume of oil is drawn across the paper surface at different speeds (fig. 9.21). The length of the oil track produced on the paper can be used to estimate both the smoothness and the absorbency of the sheet. The same basic principle is applied in a method of measuring smoothness using the IGT Printability Tester (plate 15). A fixed volume of liquid is squeezed out between the printing disc and the paper held on the impression sector of the tester. The smoothness of the paper is related to the area of the stain produced on its surface. This very versatile instrument can

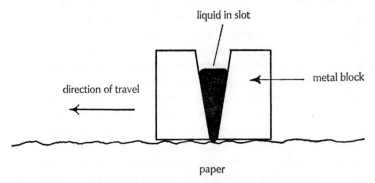

liquid in slot

metal block

direction of travel

paper

9.21 Principle of PIRA oil track method for measuring smoothness and absorbency.

also be used to measure smoothness by another method. If the ink film thickness is gradually increased while maintaining a standard printing pressure, smoothness is given by the minimum ink film thickness necessary to produce a continuous print.

No one of these methods of measuring smoothness can be said to meet all requirements. Most of them suffer from the disadvantage that they test paper in a static condition, when on the printing machine we are concerned with a dynamic situation; some involve the use of relatively expensive instruments; some require skilled operation; the time taken for some tests is excessive for control purposes. In selecting a test many factors have to be taken into account.

Absorbency

Paper is a relatively absorbent material because in between the intermeshed fibres and mineral particles are thousands of small cavities or pores which together form a maze of narrow channels capable of drawing in liquid by the action of surface tension. The absorbency or the rate at which liquid will penetrate depends on the number and size of the pores and on the properties of the liquid.

It must be clearly understood that oil absorbency and water absorbency are quite different properties. Of the two properties, the printer is far more concerned with oil absorbency, since most printing inks are oil-based materials.

Oil absorbency

By his choice of raw materials and processing conditions the papermaker can vary oil absorbency from the extremely absorbent antique paper to the virtually non-absorbent greaseproof. Open papers with large pores may absorb pigment particles along with the oily vehicle whereas with a coated paper, the large number of very small pores may only pull in the vehicle.

Since the final appearance of a print depends to a large extent on the rate of penetration of the ink, oil absorbency must rank as one of the most important printing properties of paper. Certainly many printing faults can be traced back to variations in oil absorbency. If the oil absorbency is too low, the ink will tend to lie on the surface of the paper and may set-off on to the underside of the next sheet in the stack. If the rate of oil absorbency is too high and there is excessive penetration, this may result in 'show-through' or 'strike-through' (page 120) or in a serious loss of gloss. In the case of coated papers, this rapid penetration may cause 'powdering' or 'chalking', due to pigment being left on the surface unprotected by the vehicle.

One simple method of measuring oil absorbency is to spread a thick film of oil on to the paper surface for a fixed time, and then assess the amount that has been absorbed. This is the basis of the K and N test in which a thick layer of a special ink is wiped off the surface after two minutes. Since the oil is dyed and not pigmented like an ordinary ink, the intensity of the stain left on the paper is a measure of the amount of oil absorbed. The K and N print may be visually assessed against a standard or alternatively reflectance measurements may be made. Unfortunately, variations in oil absorbency can occur across a sheet of paper, giving the final print a mottled appearance. This unevenness of absorbency will be shown up by a K and N print having dark and light patches. Because of its speed and simplicity, the test is widely used for routine control work. The K and N method of measuring ink absorbency is described in BS 4574: 1970.

The PIRA Surface Oil Absorption Tester consists of three elements: a micro-burette containing medicinal paraffin of known viscosity, a brass roller and a 1 in 20 ramp covered with a rubber offset blanket (fig. 9.22). A drop of oil of known weight is allowed to fall on to the brass roller which is then released so that it runs down the ramp. During its first revolution, the oil drop on the roller comes into contact with the rubber blanket and is spread and split between the two surfaces. During its second revolution the roller carries the thin film of oil on to the surface of the paper sample, put in its path. A stop-watch is started as soon as the oil touches the paper and is stopped when 75% of the surface film has been absorbed. If the surface of the paper is held at a low angle to the light, it is fairly easy to see the glossy film of oil gradually disappearing. 75% absorption is preferred to 100% for the end-point, since the complete disappearance of the film is often abnormally de-

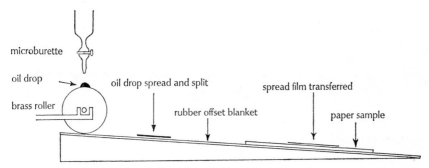

9.22 PIRA surface oil absorption tester.

layed by small areas of the surface which resist absorption. Although the test and the apparatus are simple and direct, the method does have a number of drawbacks. Very little information is obtained on the unevenness of oil absorption across a sheet. The test is very sensitive to temperature variation and should be carried out at 20°C. The interpretation of the end-point is a subjective matter, although it is possible to use a gloss meter and so remove the judgement factor. The reproducibility of the method is not good, but it does provide a satifactory means of ranking different papers. PIRA have suggested that for a coated paper, powdering will occur if the SOAT time is below 10 seconds and that for all stocks the SOAT figure should be above 30 seconds where the paper is being varnished, bronze or gloss printed and below 150 seconds to avoid set-off problems.

The PIRA method of measuring smoothness and absorbency by drawing a slot of oil across the paper at various speeds has already been referred to on page 137. A further method using the IGT printability apparatus has the advantage of applying the oil at speeds similar to those on an actual press. A drop of dyed oil is placed on the printing disc and squeezed against the paper sample, clamped round the falling segment. The length of the stain is inversely proportional to the oil absorbency of the paper.

Water absorbency

Although this property is of less interest to the printer than oil absorbency,

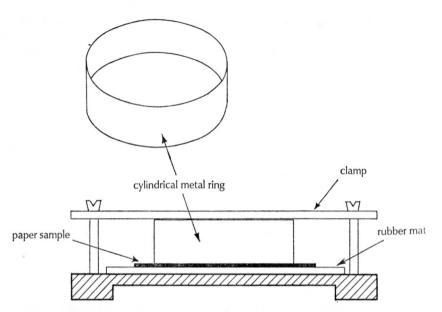

9.23. Cobb sizing tester.

the rate of water penetration can be important if a paper is going to be treated with a water-based coating like gum. The Cobb test provides a simple and effective means of measuring the amount of water absorbed into the surface of a paper. A cylindrical ring of metal is clamped firmly on top of a small weighed sample of paper, placed on a rubber mat covering a metal plate (fig. 9.23). Distilled water is poured into the circular trough formed by the metal ring so that the entire paper surface is covered to a depth of 1cm. After a suitable period of time (*eg* 60 seconds) the water is poured out of the cylinder, the paper quickly removed and its surface sponged free of drops of water with blotting paper. The sample is then reweighed, and the surface absorbing capacity of the paper is given by

$$10,000 \times \frac{\text{gain in weight (g)}}{\text{circular test area (cm}^2)}$$

The Cobb test has been accepted as a standard method and the dimensions of the apparatus and other details can be found in BS 2644:1955.

The Cobb method measures surface absorbency, but a variety of simple methods have been suggested for measuring another related property, namely, the time taken for water to penetrate right through a paper or board. These tests may be carried out as a means of assessing the degree of sizing or they may be used as a guide to the performance of wrapping paper which will meet wet conditions. In one such method, grains of a fluorescent dye mixture (1 part fluorescein to 100 parts icing sugar) are shaken into a small 'boat' of paper, floating on a trough of water, viewed under a UV lamp. The time is measured from the moment the boat is launched to the point when the moisture penetrating the sheet suddenly transforms the dull specks of powder into bright fluorescent pin points.

pH

The relative acidity or alkalinity of a paper is of considerable importance to the printer. Although we speak of a paper's pH, strictly speaking, since pH is a property of aqueous solutions, paper cannot be said to have a pH. What is really meant is the pH of an aqueous extract of the paper, the extraction having been carried out under standard conditions.

The pH of paper has been shown to be one of the factors influencing the

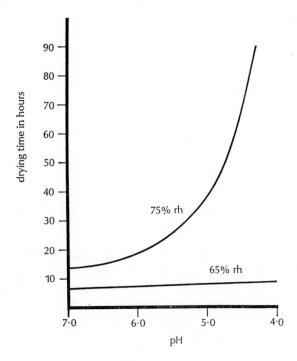

9.24 Influence of the pH of paper on ink drying (PIRA).

rate of drying of a letterpress or litho ink. The lower the pH, the longer the ink takes to dry and at pH values below 5, the drying may be seriously retarded or even stopped altogether. The problem is most acute when a paper with a low pH is printed in an atmosphere with a high relative humidity (fig. 9.24). Since humidities tend to be higher in litho than in letterpress printing, pH is a more critical factor in the litho process.

There are a number of other reasons why the pH of a paper or board should be held within certain limits. Some printing ink pigments are unstable to strongly acid or alkaline conditions, and these conditions in a paper may cause a particular colour to fade out of a print. Another example of the effect of a paper's pH is that the tarnishing of bronze powder by sulphides in the paper is much more likely to take place at a low pH.

Most coated papers have a pH above 7, due to the alkalinity of the coating. Papers with pH values exceeding 10·5 may cause fading. Uncoated papers generally have a pH of between 5·5 and 7·0. Where the figure is below 5, there is a risk of drying problems, particularly at high relative humidities.

A standard method of preparing a hot aqueous extract of paper for pH measurement is described in BS 2924:1968. This recommends that a small sample of paper together with some preheated distilled water is added to a flask (100ml of water for every 2g of paper) and the mixture boiled gently for one hour. The flask should be fitted with an air condenser so that the volume of water is maintained. After the extraction, the solution should be quickly cooled and its pH measured by means of a pH meter (as specified in BS 2924:1968) or by one of the methods involving the use of indicators.

In a pH meter, the solution to be measured is made part of a primary cell. Any cell produces a voltage which depends partly on the nature of the solution connecting the electrodes of the cell. A cell for measuring pH is arranged so that the voltage developed changes with the concentration of hydrogen ions in this solution. pH meters provide the most accurate way of measuring pH but they are relatively expensive instruments which require regular checking and careful handling in order to give consistent results.

Indicators are simply dyes which change their colour when the pH of the solution in which they are dissolved is changed. Indicator methods are simple to carry out, do not involve expensive equipment and yield results which are sufficiently accurate for many purposes. One of the more popular of these methods uses the Lovibond Comparator, a small plastic box in which the colour of the test solution containing a certain indicator is visually compared with permanent colour glasses set into a circular disc. These glasses match the colours which that same indicator gives in solutions of known pH.

A more approximate guide to the pH of a paper can be obtained by smearing a drop of indicator over it with a clean piece of glass, and comparing the colour obtained with the colour of a series of specially prepared slides developed by PIRA.

Tests for the constituents of paper

A wide range of tests are available to detect the presence of and, in some cases, find the amount of the various materials that may be present in paper. A survey of these tests is beyond the scope of this book but those seeking further information are referred to the appropriate chapter in the well-known book by Dr Julius Grant, *A Laboratory Handbook of Pulp and Paper Manufacture*, published by Arnold. However, mention must be made of some of these tests that are of particular interest to the printer.

Cellulose fibres

We have already seen that the various types of cellulose fibre used in making paper can be recognised by their appearance under the microscope (Chapter 5). In preparing a slide for the microscope, a sample of the paper about 5 cm square is torn up into small pieces and disintegrated in a relatively large quantity of distilled water. In effect, one is trying to reverse the action of the paper machine and produce a dilute suspension of separated fibres. The process of disintegration may be speeded up by boiling the torn pieces in water or where this is not effective in a 0·5% solution of sodium hydroxide. When sodium hydroxide has to be used, the fibres should be thoroughly washed afterwards. Chewing a small piece of paper provides a rapid and convenient method of breaking down the paper structure before separating the fibres in water, but it is not recommended! A small amount of the fibre suspension is poured out on to a clean microscope slide, which is then placed carefully on a hot plate at about 70°C in order to remove the water.

The identification of fibres under the microscope is made easier by the use of stains, which selectively colour the different types of fibre. For example, the general-purpose Herzberg stain will colour rag fibres red, mechanical wood yellow, esparto violet, and chemical woods blue. The appearance under the microscope of some of the more common fibres is shown in plates 1–6. Enlarged photomicrographs and a very useful introduction to paper microscopy may be found in F D Armitage's *Atlas of Paper Fibres*.

Stains may also be applied to a sheet of paper to indicate the presence of certain fibres. Two drops of a solution of phloroglucinol followed by a drop of concentrated hydrochloric acid will quickly show the presence of mechanical wood fibres by turning red.

Loadings

Information on the amount and type of loading can be gained by igniting the paper and examining the residue. A standard method for determining the ash content of a paper is set out in BS 3631:1963. Normally about 1g of the sample is accurately weighed in a crucible, which is then ignited in a furnace at 850°C until the ash is a greyish white. The crucible must be cooled in a desiccator to avoid the intake of moisture before being accurately weighed a second time. A simple calculation is necessary to convert the ash content into the actual percentage of loading, because of losses in weight due to chemical changes and moisture lost from the original loading. For example, china clay loses approximately 14% by weight in the ashing of a paper. If the nature of the filler is not known, a chemical analysis may be carried out on the ash.

Reducible sulphur

It has been shown that the tarnishing of bronze and gold inks can be due to the presence in the paper of sulphur compounds that are readily reduced to hydrogen sulphide. The test for reducible sulphur consists basically of reducing these compounds to hydrogen sulphide and then passing this gas over lead acetate paper. The hydrogen sulphide reacts with the lead acetate, forming brown lead sulphide, and the intensity of the stain on the paper can be compared with stains produced by known quantities of sulphur under similar conditions. The result of this test must be considered together with the pH of the paper since the tolerable concentration of reducible sulphur is reduced at a low pH. Details of a method of determining the concentration of reducible sulphur in paper and board may be found in the BPBMA Proposed procedure No 49.

Printing tests

While it would be true to say that each of the tests so far described has some value in assessing the printability of a paper or board, it must be admitted that in many cases this value is extremely limited and that perhaps the only way of reasonably predicting the printing performance of a paper is to print it. This approach to printability tests is so obvious that it is often overlooked. Sometimes it is possible to insert sample sheets of a newly delivered paper or board into the current production run of the same or a similar job. Usually it is not practicable to run the paper on a production machine, but this does not rule out the possibility of printing the paper by some other means. For example, the printing might be done on a small training machine or even on a proof press, but it should be stressed that printing tests of this kind are only of value when both the printing conditions and the method of assessing the print have been standardised. It is certainly not easy to establish reproducible conditions on a commercial printing machine and it is this difficulty that highlights the value of the printability testers, which bring printing into the laboratory. These instruments do make it possible to print samples of a paper under standard conditions, and the most successful of these, the printability tester developed by IGT, has already been referred to on a number of occasions. Other instruments which can be used for making standard prints and for assessing the printability of paper are listed on page 256.

10. Printing ink – drying methods

The function of a printing ink is to form a permanent coloured image on paper or some other substrate. Most printing inks can be considered as having two main components, firstly a *pigment*, which gives the image its contrast against the background of the paper, and secondly a *vehicle*, which carries the pigment to the paper, provides adhesion to the paper surface and protects the image during the lifetime of the print.

In all the major printing processes, printing ink is applied to the substrate as a liquid, which is then converted to a solid as soon as possible after impression. Although this has always been the accepted method of producing a permanent ink film, it should be borne in mind that it is perfectly possible to transfer a printing ink in powder form; in fact, this method is used in certain electrostatic printing processes now being developed. After being applied, the powdered ink is quickly heated and cooled, so that it is firmly fused on to the paper surface as a dry tough film.

Whether a printing ink is applied in liquid or powder form, the resulting dry ink film must have the hardness, adhesion and flexibility to allow it to withstand handling during its lifetime. These properties are achieved by surrounding the pigment particles in the ink with a resinous polymeric material. It follows that an ingredient common to all ink vehicles is a resin of high molecular weight or at least a substance which will develop these resinous properties after impression. Vehicles for news inks provide an exception for they often contain no resin, and anyone who has blackened his hands reading a newspaper will agree that the black pigment is not well protected.

Ink drying methods

The application of a surface coating, whether it be an ink, a paint or some other material, normally involves the change from the liquid to the solid state, a process which is loosely known as *drying*. Ideally, a printing ink should be completely stable in the can and on the press, but on striking the surface of paper or some other substrate it should be instantly converted into a tough

146

solid film. In practice this happy situation is never entirely achieved and because of other requirements the formulation of a printing ink usually involves a compromise. It is important to remember that printed ink films are very thin, usually much thinner than the substrate on which they have been printed.

Various physical and chemical methods are available to bring about the drying of a printing ink and these methods will now be discussed. Drying is often achieved by combining two or more of these methods.

OXIDATION AND POLYMERISATION

When certain oils are left exposed to the atmosphere, they gradually thicken and eventually become solid materials. Linseed oil is one such *drying oil,* which has been used in making inks from the earliest days of printing. The chemical reaction which brings about the change of state of a drying oil is a combination of *oxidation* and *polymerisation.*

The nature of polymerisation in which small molecules (monomers, normally gases or liquids) link up to form giant chains or networks of molecules (polymers, normally solids) is discussed in Chapter 4. The process in which linseed oil is converted from a liquid to a solid is broadly similar, except that oxygen from the atmosphere plays an essential part in the reaction. In its natural state, linseed oil dries too slowly and is not sufficiently viscous to be suitable as a printing ink vehicle. However, these disadvantages are overcome by various heat treatments, some of which are carried out under vacuum. The drying oil molecules link up into longer chains and so become more resistant to flow. The degree of polymerisation depends on the length of time for which the heating is continued, and thus the more viscous lithographic varnishes are the result of more prolonged heating than a medium or tint varnish. Heat treated drying oils of this type have been the traditional vehicles for letterpress and lithographic inks for many years. In more recent times, the natural drying oils like linseed oil have been chemically modified in various ways and completely synthetic drying oil vehicles developed in order to produce quick drying inks with improved performance.

When a drying oil varnish is exposed to the atmosphere the process of linkage between molecules continues until a hard, flexible and transparent material is obtained. As is indicated in a later chapter (page 165), the molecules of all drying oils have carbon chains which include points of unsaturation, that is, neighbouring carbon atoms each having less than the maximum possible number of hydrogen atoms attached (fig. 10.1). The reactions between drying oil molecules and oxygen take place at these unsaturated points on the carbon chains, enabling bridging links to be built across to neighbouring molecules.

Printing inks and paints drying by oxidation and polymerisation always contain a small proportion of *driers*, materials which greatly speed up the rate of drying. These driers act as *catalysts*, that is substances which assist a chemical change without undergoing any permanent change themselves. Printing ink driers are metallic soaps which act as oxygen carriers, transferring atmospheric oxygen to the drying oil molecules faster than they can obtain it for themselves.

$$-\overset{\displaystyle H}{\underset{\displaystyle H}{C}}-\overset{\displaystyle H}{\underset{\displaystyle H}{C}}-\overset{\displaystyle H}{\underset{\displaystyle H}{C}}-\overset{\displaystyle H}{\underset{\displaystyle H}{C}}-\overset{\displaystyle H}{\underset{\displaystyle H}{C}}-$$

saturation

$$-\overset{\displaystyle H}{\underset{\displaystyle H}{C}}-\overset{\displaystyle H}{C}=\overset{\displaystyle H}{C}-\overset{\displaystyle H}{\underset{\displaystyle H}{C}}-$$

unsaturation

10.1 Saturated and unsaturated regions of carbon chains.

We have already seen that the drying of a letterpress or lithographic ink can be seriously retarded if the pH of the paper is too low, particularly if the relative humidity of the pressroom is high. This has been shown to be due to a chemical reaction between the drier and the acidic component in the paper, so leaving the ink film with an insufficient concentration of drier. Oxidation and polymerisation drying can also be retarded by an inadequate supply of another essential component in the reaction, oxygen. These conditions may arise when printed sheets are held very close together in a stack, so that air is unable to flow between the sheets.

ABSORPTION

Whenever a liquid comes into contact with a porous material, surface tension will cause some of that liquid to be drawn into the openings in the surface structure. The network of narrow channels between fibres and particles of mineral loading make most papers extremely porous materials, and the capillary attraction of an ink into a paper surface provides the simplest physical method of drying. Strictly speaking the liquid is not being converted into a solid, but since it becomes an integral part of the solid paper, the result

Printing ink – drying methods

ink transferred

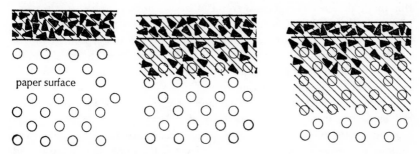

paper surface

10.2 Stages in absorption drying.

is much the same. News inks, some other letterpress inks and certain web-offset inks applied to absorbent stock, dry solely by penetration. Combined with other methods, absorption makes an important contribution to the drying of all printing inks, except when non-absorbent materials like metallic foils, tin plate, plastic films and certain papers are being printed.

The penetration of ink into paper has been shown to occur in two stages, a rapid penetration under pressure at the moment of impression followed by a slower penetration due to capillary forces. In this second stage, some separation of the vehicle from the pigment may take place (fig. 10.2). It is probable that the separate absorption of the vehicle into the paper proceeds until the particles of pigment in the ink come close enough together for the spaces between them to exert an equal pull on the liquid.

Absorption drying can only be carried out successfully if there is a proper balance between the oil absorbency of the paper and the viscosity of the ink. If there is insufficient penetration, the ink left on the surface of the paper may not be able to dry quickly enough to avoid set-off on to the next sheet. In extreme cases this ink can act as an adhesive between the printed surface and the back of the next sheet causing the two to stick together, *ie* block. On the other hand, if too much of the vehicle is absorbed into the paper, unprotected pigment is left on the surface and this may easily rub off or set off on to the following sheet. This effect, known as *powdering*, is more common with coated papers than it is with papers having a more open surface, because the latter allow a larger proportion of the pigment to be carried into the structure of the paper with the vehicle. The excessive absorption of a vehicle can also lead to show-through or strike-through, making the printed image visible from the other side of the sheet (page 120).

The great advantage of absorption drying is that it takes place very quickly, and the highest press speeds have been achieved with inks that rely solely on this method. Since the general trend in every modern industry is towards higher productivity and faster-running machines, absorption drying appears to provide an attractive means of coping with higher press speeds. Unfortunately a print produced with an ink which has largely been absorbed into the paper is bound to have a dull and lifeless appearance because very little light can be specularly reflected from its surface. For example, a standard news ink, consisting basically of a dispersion of carbon black in a mineral oil, strikes rapidly into the paper on impression leaving a thin layer of oil and pigment on the surface, with a matt appearance. In order to produce printing with an attractive glossy appearance and yet at the same time retain the advantage of rapid drying by absorption, '*quicksetting*' inks have been developed using a two-phase vehicle. The two components of the vehicle are

(1) a resin usually combined with a drying oil, thinned with
(2) a thin high-boiling solvent, usually a paraffin hydrocarbon distillate, boiling around 300°C.

When an ink of this type is printed on to paper the thin component is quickly absorbed leaving the remainder of the ink on the surface. This viscous mixture of resin, drying oil and pigment sets and then is slowly converted into a hard dry film as a result of oxidation and polymerisation.

EVAPORATION

Many of the everyday uses of the word 'drying' refer to drying by the evaporation of water, for example, from the wet sail of a dinghy or from a line of wet clothes. In a somewhat similar way, the evaporation drying of a printing ink comes about by the removal of a solvent or a mixture of solvents into the atmosphere, but in most cases water is not used as a solvent.

In the simplest case, a printing ink drying solely by evaporation has a vehicle consisting of a hard resin dissolved in a volatile solvent. After printing, the solvent evaporates leaving the resin binding the pigment to the surface of the paper. The removal of the solvent may be speeded up with the aid of heat, blown air, etc. Evaporation is the most important method in the drying of gravure and flexographic inks, and it may be used as the only means of drying these inks when they are applied to non-absorbent materials like aluminium foil and plastic films.

A wide range of solvents are available for use in evaporation inks and the choice of solvent has an important effect on the properties of the ink. Obviously the solvent selected must effectively dissolve a suitable resin binder but it must also lead to an ink with reasonable press stability and a satisfactory rate of drying. One of the main factors influencing the rate of evaporation is the *vapour pressure* of the solvent.

There is always a tendency for some molecules in the surface of a liquid to escape into the space above, although many molecules fall back into the liquid. As the temperature of the liquid is raised the molecules become more agitated and an increasing number are moving fast enough to escape from the surface. At any given temperature an equilibrium concentration of molecules is formed in the closed space, and this 'traffic' of escaping molecules gives rise to the vapour pressure of the liquid. When the temperature of a liquid is raised to a point when its vapour pressure just exceeds the external atmospheric pressure the liquid will 'boil', and molecules leave the surface continuously. As a rough guide, the lower the boiling point the faster a solvent will evaporate and the less additional heat will be required to accelerate drying. To be more precise, the rate of evaporation of a solvent depends on its vapour pressure at the temperature of drying rather than at its boiling temperature, illustrated by the fact that toluene (boiling point 111°C) evaporates more rapidly than water (boiling point 100°C) at room temperature. Most volatile solvents are highly inflammable and so safety precautions often prevent the use of the fastest drying solvents because their *flash point* is too low. The flash point is the lowest temperature at which a solvent will form a vapour/air mixture which will produce a flash, *ie* explode weakly, without burning continuously. It should be mentioned that quite apart from the safety factor, very rapid rates of evaporation prevent a surface coating from flowing out properly and can have other adverse effects on the film structure of a surface coating. Other factors which may influence the choice of solvents are their odour, in the case of inks for packaging, the nature of the printing plate, which may be attacked by some solvents, and the nature of the material being printed, since with some plastic films it is necessary to include a solvent which will slightly dissolve the surface of the film to achieve an adequate bonding of the ink.

PRECIPITATION

Although less important than the methods discussed so far, precipitation does provide another means of rapidly converting a liquid surface coating into a solid, one which has been successfully used commercially in letterpress *moisture-set* inks. If an ink with a vehicle consisting of a water-insoluble resin dissolved in a water-miscible solvent is printed on to paper, water from the paper and the surrounding air dilutes the solvent to the point where it can no longer hold the resin in solution, and this resin is precipitated round the pigment to form a dry film of ink. If the amount of water in the paper is insufficient to cause rapid precipitation, a jet of steam may be blown on to the wet print. The solvents used in moisture or steam-set inks are invariably *glycols* since these materials are miscible with water, hygroscopic and have good solvent power for a reasonable range of resins.

The great advantage of these glycol-based inks is that they are free from the unpleasant odours of the by-products formed during the polymerisation of drying oils. On the other hand, one great disadvantage is their poor press stability due to the tendency for them to absorb water from the atmosphere, bringing the resin out of solution on the distributing rollers instead of on the paper. Because of this problem the trend in recent years has been towards glycol-based inks containing enough alkali to make them more stable to the absorption of moisture, a move away from the moisture-set principle. Great care must be taken in selecting suitable rollers to carry inks of this type,

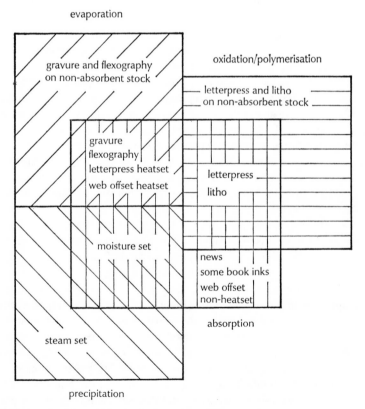

10.3 Drying methods in combination in the major printing processes.

because glycols strongly attack the traditional gelatin 'compo' roller. Glycol-based inks are not suitable for lithography owing to the presence of the fountain solution but they are being used in dry offset printing.

Other drying methods

Oxidation and polymerisation, absorption, evaporation and precipitation are the drying methods in most general use, but there are other methods which find a specialised application. Most carbon papers are produced with an ink which is solid at room temperature, but which melts in the heated duct of the printing machine and is applied from a heated letterpress surface or heated gravure cylinder, to dry by solidification when it cools on the surface of the paper. Some vehicles for inks and coating lacquers contain two components which chemically crosslink when the prints are subsequently heated. More recently there have been important developments in ink systems in which drying is brought about by other forms of radiation.

Drying units which emit microwave radiation were first used commercially on web-offset newspaper presses. The microwaves penetrate the ink film and drive off the solvents without intense direct surface heat. The efficiency of microwave drying greatly depends on the formulation of the ink. The very promising photo-reactive inks dry as a result of the polymerisation of a photo-sensitive vehicle on exposure to ultra violet radiation. Although more expensive than their heatset and quickset counterparts, inks drying by photo-polymerisation are potentially attractive since they are free from odour and from any problem of air pollution, a factor of growing importance. Further into the future, electron beam curing in which inks are dried by irradiation with an electron beam, is based on a similar type of polymerisation reaction.

Drying methods in combination

We have seen that some inks rely solely on one method of drying. When the material being printed is completely non-absorbent, gravure and flexographic inks dry only by evaporation, whilst litho and most letterpress inks would dry only by oxidation and polymerisation. In many more cases, drying takes place by a combination of drying methods and when paper or board is printed, absorbency always makes some contribution to the drying process. The ways in which drying methods are combined in the major printing processes are illustrated diagrammatically in fig. 10.3.

11. Printing ink – pigments

The colouring matters used in printing inks are generally *pigments* rather than *dyes*. Pigments are finely divided coloured substances which are insoluble in the vehicle, and so they are dispersed rather than dissolved. Dyes, which are soluble in the vehicle, are only used for certain specialised applications as for example in flexographic inks and for correcting the tone of a black newsink.

Many thousands of different pigments exist but relatively few of these have the combination of properties required in a printing ink. The more important of these properties are good colour strength, reasonable stability to light and chemicals, fine particle size and the ability to be dispersed in normal ink vehicles to give inks with good flow properties. The colouring matter in a particular ink may also need to have a number of special properties in order to meet the demands made on the ink, by the printing process, by subsequent processing, and during the working life of the final print.

Clearly in selecting pigments for inks, the ink maker needs to have a great deal of information on their properties. The range of this information is well illustrated by the product information books published by the pigment manufacturers. One such book includes the following data on each pigment:

(1) The chemical constitution of the pigment and a reference to its position in the Colour Index, which gives classified information on all colouring matters:

(2) Colour prints of the pigment at full strength and in reductions with alumina and titanium dioxide.

(3) Colour measurement data: CIE co-ordinates.
Spectrophotometric curve.

(4) Pigment properties: Specific gravity.
Bulking volume (litres/kg).
Oil absorption (g oil/100g pigment)
pH of 2% aqueous extract.
Fastness to solvents, plasticisers
and oils (*eg* water, ethanol, xylene,
dibutyl phthalate, linseed oil).

(5) Application properties:	Concentration to give standard flow.
	Relative colour strength.
	Dispersibility.
	Transparency.
	Print finish.
	Lithographic breakdown.
(6) Fastness of print:	Heat and sterilisation at various temperatures.
	Light.
	Reagents (water, acid, alkali, detergent, soap, spirit varnish, wax, butter, coconut oil).

The prints of inks based on the pigment are of very practical value, although it must be remembered that since the shade produced by a pigment depends on the thickness of the ink film, the type and quality of the extender used, the type of vehicle used and on the quality of the paper, it is difficult to define this shade with any degree of precision. The prints are normally made at full strength and also in reductions, when the pigment is mixed with white pigments or extenders *eg* 5 parts of alumina to 1 part of pigment. This visual information on the colour of the pigment may be supported by CIE data and spectrophotometric curves obtained from colour measuring instruments.

The *specific gravities* of pigments vary widely and table 11.1 compares the specific gravities of two typical organic pigments with those of two inorganic pigments. The *bulking volume* gives an indication of the volume occupied by equal weights of different pigments. Those pigments with a high specific gravity have a low bulking volume. The figure for *oil absorption* is the weight in grammes of linseed oil required to give a coherent paste with 100 g of pigment. The maximum amount of pigment that can be dispersed into a vehicle to produce a handleable paste varies considerably and the *percentage concentration to give standard flow* is one way of comparing pigments in this respect. This percentage concentration is low for those pigments with a high oil absorbency (Table 11.1). In turn, a high oil absorbency is usually associated with pigments having a small particle size and hence a large surface area to be wetted by the oil. The figure for *dispersibility* indicates the ease with which aggregates of pigment are broken down during milling. Each of these properties relate particularly to the performance of the pigment during ink manufacture. Of the other properties listed, some are of special value in judging the suitability of the pigment for a particular printing process (*eg* litho breakdown,

Table 11.1 Properties of some typical inorganic and organic pigments

	Specific gravity	Bulking volume litres/kg	Oil absorption g oil/100g pigment	Concentration to give standard flow %
Middle Chrome	6·2	1·1	10	68
Scarlet Chrome	5·4	0·96	12	70
Hansa Yellow	1·4	6·3	44	15
Benzidine Yellow	1·3	4·4	57	10

colour strength, water resistance, heat resistance), others may have a bearing on subsequent processing (eg resistance to heat, wax, varnish) and any of the fastness properties (heat, light and reagents) may be relevant to the function of the final print. One factor which does not appear on the list is the pigment's cost. Clearly in many applications this can be the overriding consideration.

TYPES OF PIGMENTS

Broadly speaking pigments for printing inks can be considered under three main headings: carbon blacks; inorganic pigments including white pigments and extenders; and organic pigments.

Carbon blacks

Carbon black pigments consist largely of the element carbon, the basic constituent of all animal and vegetable materials. When many of these carbon-containing substances are burnt in air a soot is formed. If the supply of air is carefully controlled the soot can be deposited in an extremely fine form which behaves as a most effective black pigment. By varying the conditions in the reaction many different types and grades of carbon black can be produced differing in their chemical purity, their surface properties and in their particle size. For printing inks, it is normal to use grades of either furnace or channel blacks, produced by burning oil or natural gas in a limited supply of oxygen.

Carbon black must rank as the most important pigment in printing inks, since it is used to colour virtually all black inks. It is relatively cheap and yet possesses a remarkable range of properties including good colour strength and excellent resistance to light, heat, moisture and most chemicals. Carbon blacks have a much smaller particle size than any other pigment, ranging from as little as 10nm for some channel blacks to 200nm for some furnace and lamp blacks. These tiny particles present a very large surface area to be wetted by a vehicle and this largely accounts for high oil absorptions and difficulties in dispersing the pigment. Carbon black pigments also tend to absorb metallic soap driers from black letterpress and litho inks, and because of this, drier concentrations have to be higher than normal in these inks.

Inorganic pigments

The inorganic pigments used in printing inks include all the white pigments and extenders together with a small but important group of coloured substances which still play an important role in surface coatings despite the great advances made by synthetic organic colouring materials. This group includes colouring matters which occur naturally on the earth's surface like china clay, chalk and the siennas, ochres and umbers. However, where these substances are used in printing inks, synthetic grades are invariably preferred to the coarse and gritty natural products. The manufacture of these synthetic grades allows the particle size and surface properties to be carefully controlled.

White pigments and extenders

The white inorganic substances used in printing ink manufacture may either be *pigments*, like titanium dioxide and zinc oxide, or *extenders* like alumina hydrate and precipitated chalk. The essential difference between a white pigment and an extender is that while pigments contribute much to the opacity and hiding power of an ink film, extenders are relatively transparent when dispersed in an ink vehicle. An explanation of these different effects is found in the relative refractive indices of the pigment and the vehicle.

Light rays are refracted by particles that have a refractive index differing from that of the film in which they are dispersed (fig. 11.1). The greater the difference between the two refractive indices, the more this scattering of light takes place and the greater will be the opacity and hiding power of the ink film. The refractive indices of most printing ink vehicles fall within the range 1·4–1·6. Rutile titanium dioxide is the most opaque white pigment by virtue of its high refractive index (2·72). At the other extreme, alumina hydrate (refractive index 1·54) is an extremely transparent extender. Other white pigments and extenders lie between these extremes, zinc oxide showing medium opacity and gloss white, blanc fixe and china clay medium transparency. Precipitated chalk (calcium carbonate) is widely used as a general purpose transparent extender.

The crystalline form of a pigment can have an important influence on its properties, including its hue. For example titanium dioxide exists in three crystalline forms, the two forms used in printing inks being *Rutile* and *Anatase*. The rutile form is more opaque than the anatase but slightly yellower in tone.

Although extenders are relatively cheap their purpose is not simply to re-

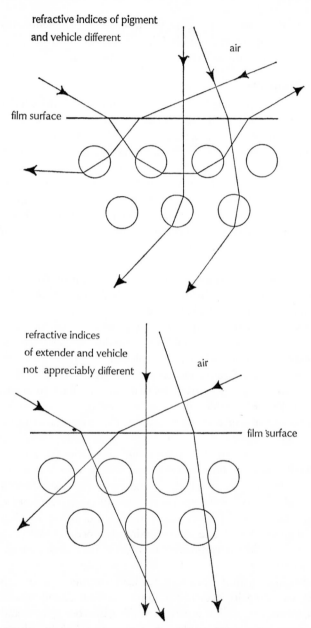

11.1 Light scattering and opacity depends on there being a difference between the refractive indices of pigment and vehicle.

duce the price of an ink. The colour strength of many pigments is so great that they frequently have to be used at well below full strength. Extenders provide a means of reducing this colour strength without introducing whiteness and opacity into the ink. They are also used to improve the consistency and general working properties of an ink.

Coloured inorganic pigments

A large number of simple inorganic oxides and salts are coloured but relatively few of them have sufficient colour strength or are stable enough to be useful as pigments. Among the few that are suitable, there are some that are of considerable importance in printing inks.

The *lead chrome* group of pigments range from a greenish yellow to a deep orange. Chemically they consist mainly of lead chromate $PbCrO_4$, which is precipitated when solutions of sodium chromate and lead nitrate or acetate are mixed. The range of hues is obtained by precipitating various proportions of other lead salts at the same time and by adjusting the reaction conditions. The lead chrome pigments are extremely opaque and so they are particularly useful when maximum hiding power is required. They are cheap, easily dispersed in printing ink vehicles and show good resistance to most chemicals. Their light fastness is also good, although they have a tendency to darken on exposure to the atmosphere. Because of their high lead content they are not recommended for use in inks for food packages. Like other inorganic pigments, lead chromes have a relatively high specific gravity and so show a tendency to settle out of liquid inks. *Molybdate orange*, a mixture of lead chromate, molybdate and sulphate, is another extremely opaque pigment widely used in many types of ink.

Of the blue inorganic pigments, the *bronze blues* are the most important. Chemically they are related to ferric ferrocyanide $Fe_4[Fe(CN)_6]_3$ and the deep blue pigments may be given a green or reddish undertone by varying the conditions of manufacture. The bronze blues have good resistance to light, heat, oils and solvents, but they are decomposed by alkalis. *Ultramarine* finds fewer applications in printing inks, due to its limited colour strength and poor texture.

The naturally occurring *ochres*, *siennas* and *umbers* are complex chemical mixtures, largely consisting of iron oxide. Their hues range from yellow to reddish-brown, but they have a dullness which is characteristic of earth pigments. Synthetic versions of these iron oxide pigments are preferred for printing inks. Although they are coarse and difficult to grind, their cheapness and natural permanence to light makes them useful in a limited range of inks.

Generally speaking inorganic pigments consist of hard dense crystals which in some cases are difficult to grind into a vehicle and which can give inks with poor working properties. They are very fast to light and in many cases to chemicals. Perhaps their greatest advantage is their low cost. Certainly, they are perfectly adequate for many types of ink and, with improvements in their texture and dispersibility made by the manufacturers in recent years, there is little doubt that they will continue to play a significant role in the colouring of printing inks.

Organic pigments

Organic colouring matters, obtained from plant and animal sources, like indigo, madder and cochineal, have been known for many hundreds of years but it was not until 1856 that the first synthetic organic colouring matter was produced. In that year, an eighteen-year-old chemistry student, William Perkin, made an unexpected discovery of a mauve dye that was to be the starting point in the development of thousands of synthetic dyes and pigments. Today these are the major materials used in the colouring of textiles, plastics and surface coatings.

11.2 Structures of aromatic hydrocarbons.

11.3 Steps in the manufacture of a simple organic pigment.

We have seen that in general, inorganic pigments are simple chemicals with small molecules, eg lead chrome $PbCrO_4$, titanium dioxide TiO_2. In contrast, organic pigments are relatively large and complicated molecules containing large numbers of atoms. In making these synthetic colours, the chemist starts with aromatic hydrocarbons obtained from crude oil or coal tar. These include benzene, toluene, napthalene and anthracene, each having a chemical structure related to benzene, with its six carbon and six hydrogen atoms linked together in the form of a hexagon (fig. 11.2). In a series of chemical reactions these hydrocarbons are modified and finally linked together to form coloured compounds with large molecules. A classification of the various classes of organic pigments by their chemical structure is beyond the scope of this book, but in order to give some idea of how they are obtained, the chemical steps involved in the manufacture of one fairly simple pigment, toluidine red, are outlined above (fig. 11.3).

Toluidine red belongs to the largest single group of synthetic organic colouring matters – the *azo* dyes and pigments. These all contain the azo group $-N=N-$, and their manufacture always involves the conversion of an aromatic amine into a diazo compound followed by coupling with a second aromatic compound, as shown in the case of toluidine red. By varying the choice of amine and coupling component, it is possible to produce many thousands of different azo colours. Azo pigments used in printing inks include the Hansa Yellows, Benzidine Yellows, Permanent Reds, Lithol Rubines and Lake Red C.

Most of the azo colours are yellows, oranges and reds, but other classes of organic pigments provide blues and greens. New synthetic pigments are still being discovered but relatively few survive the extensive programmes of development and testing which are necessary before they can become commercial products. One of the last major developments came in the 1930s when an entirely new range of blue and green pigments, the *phthalocyanines*, was discovered and later marketed by ICI under the trade name Monastral. Two of these remarkably stable colours are used in formulating the greenish blue (cyan) inks for three and four colour printing.

Fuller information on the various chemical classes of organic pigments can be found in the *Printing Ink Manual*, details of which are given on page 248.

Compared with the inorganics, organic pigments provide brighter shades and superior colour strength. Their specific gravities are much lower (see Table 11.1), so they show less tendency to settle out of an ink. Whilst inorganic pigments are inclined to have hard crystalline structures which may be difficult to grind and can cause wear of litho plate images and gravure doctor blades, organic pigments have a soft texture and are free from problems of wear. Organic pigments are generally transparent, whereas most inorganics are opaque. The resistance of organic pigments to light, heat and chemicals varies a great deal. Some of them are quite weak in this respect and may have poor soap or oil resistance, a tendency to bleed in solvents or only moderate stability to light and heat.

Clearly the pigment that is good in every respect does not exist. All pigments, whether they are organic or inorganic show some weakness. For this reason, the ink maker has to select a pigment for a particular ink with great care. It is in the printer's best interests to give his ink supplier the maximum information on the nature of the job, including the printing process, the type of machine, the material being printed and the demands made on the ink by subsequent processing and finally as part of a printed product.

12. Printing ink – vehicles

We have seen in an earlier chapter that printing inks basically consist of a colouring matter dispersed in a liquid vehicle, and that after printing, various drying methods are used to convert this liquid mixture into a solid film of ink on the paper. In practice, the number of separate ingredients in an ink is always greater than two and in some cases may exceed ten. This is partly because it is often necessary to include several pigments to achieve a particular shade, and also because the vehicle, which makes up the rest of the ink, contains a number of different types of material, coming from the following main groups of substances: oils; resins; solvents; plasticisers; and driers.

OILS

Oils may be 'animal, vegetable or mineral'. They range from animal oils or fats like lard and vegetable oils like linseed oil or olive oil to the mineral oils obtained from petroleum. Some oils 'dry' when they are exposed to the atmosphere in a thin film, and these *drying oils* are the basic liquid components in letterpress and lithographic inks.

Drying oils

Most drying oils are obtained from the seeds of plants. The one most used in surface coatings, linseed oil, is extracted from the seeds of the flax plant, whilst tung oil and oiticica oil are obtained from the nuts of certain trees. Like most natural products, drying oils are mixtures of several substances, but they are largely made up of *triglycerides* or tri-esters of glycerol (glycerine).

An ester is formed when an alcohol reacts with an organic acid. For example, methyl alcohol will react with acetic acid to form the ester methyl acetate.

$$CH_3OH + CH_3.COOH \longrightarrow CH_3.OOC.CH_3 + H_2O$$
| Methyl | Acetic | Methyl | Water |
| alcohol | acid | acetate | |

The reactive groups are the hydroxyl (OH) group in the alcohol and the hydrogen (H) in the acetic acid. These combine to form a molecule of water and the methyl and acetate groups come together to form the ester. The general equation for this type of reaction may be written:

$$R.OH + R.'COOH \longrightarrow R.OOC.R' + H_2O$$

alcohol   acid   ester   water

If the alcohol has two hydroxyl groups available instead of only one, then it is possible for it to react with *two* molecules of acid to form a diester.

$$R\begin{cases} OH \\ OH \end{cases} + 2.R.'COOH \longrightarrow R\begin{cases} OOCR' \\ OOCR' \end{cases} + 2H_2O$$

Glycerol has *three* hydroxyl groups available and so it is able to form triesters, *eg* glycerol triacetate.*

$$\begin{array}{l} CH_2.OH \\ | \\ CH.OH \\ | \\ CH_2.OH \end{array} + 3.CH_3.COOH \longrightarrow \begin{array}{l} CH_2.OOC.CH_3 \\ | \\ CH.OOC.CH_3 \\ | \\ CH_2.OOC.CH_3 \end{array} + 3H_2O$$

glycerol   acetic acid   glycerol   water
                                          triacetate

The glycerol tri-esters which form the bulk of linseed oil consist of glycerol combined with a mixture of certain fatty acids. These fatty acids have the same carboxylic acid group as acetic acid,

$$\text{Acetic acid} \quad CH_3 - \begin{bmatrix} C \nearrow^{O} \searrow_{OH} \end{bmatrix}$$

carboxylic acid
group

but in place of the $-CH_3$ group they have long carbon chains with hydrogen atoms attached. The fatty acids which predominate in linseed oil are linolenic acid (48–61% of total acids), oleic acid (20–25%) and linoleic acid (about 15%).

* When both alcohol and acid have two or more reactive groups a polyester will be formed (page 46).

Oleic $CH_3.(CH_2)_7.CH = CH.(CH_2)_7.COOH$
Linoleic $CH_3.(CH_2)_4.CH = CH.CH_2.CH = CH.(CH_2)_7.COOH$
Linolenic $CH_3.CH_2.CH = CH.CH_2.CH = CH.CH_2.CH = CH.(CH_2)_7.COOH$

Notice that each of these acids has a linear chain of 18 carbon atoms, with positions on these chains where neighbouring carbon atoms have less than the maximum possible number of hydrogen atoms attached, *ie* points of unsaturation (page 148).

The structural formula of a triglyceride in linseed oil may be written

$$CH_2.OOC.R_1$$
$$|$$
$$CH.OOC.R_2$$
$$|$$
$$CH_2.OOC.R_3$$

where R_1, R_2 and R_3 represent the long hydrocarbon chains of the fatty acids, *eg* R_1 might be the hydrocarbon chain from linolenic acid

$$CH_3.CH_2.CH = CH.CH_2.CH = CH.CH_2.CH = CH.(CH_2)_7{-}$$

The real structure of a triglyceride can only be properly shown in a three-dimensional model, but it can be formally represented as shown in fig. 12.1, where the straight lines represent the glycerol group and the wavy lines the fatty acid chains.

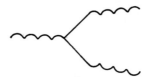

12.1 Diagrammatic representation of triglyceride molecule.

The drying properties of linseed oil and similar oils depend on the unsaturated portions of the fatty acid chains, the double bonds $-CH{=}CH-$ providing points for attack by atmospheric oxygen and sites for bridging linkages between triglycerides in the polymerisation process. This linking up of triglyceride molecules in the drying process is represented diagrammatically in fig. 12.2.

Linseed oil

If a thin layer of raw linseed oil is spread out on a sheet of glass and left exposed to the air, it increases in weight and eventually becomes a dry film.

12.2

The increase in weight is due to the uptake of atmospheric oxygen in the drying process. If the sheet of glass is kept in an atmosphere of nitrogen in a closed container, there is no increase in weight and no drying takes place.

A layer of raw linseed oil, exposed to the atmosphere at normal temperatures, may take up to four days to dry to a hard film. Thicker oils, which dry much more rapidly, are produced by heat-treating the raw oil and it is these heat-treated linseed oils which have been the major components in letterpress and lithographic inks from the earliest days of these printing processes.

Litho varnishes or stand oils are made by the prolonged heating of alkali-refined linseed oil at a temperature of 270–290°C. Under these conditions reactions occur at the double bonds on the fatty acid chains and linkages are formed between neighbouring molecules of triglycerides. As the polymerisation proceeds the viscosity of the oil increases and by stopping the process at various stages, a range of litho varnishes with different viscosities can be produced. Other partially polymerised forms of linseed oil are the *Boiled oils*, produced by heating raw linseed oil at 90–150°C in the presence of air and driers, and *Blown oil* made by blowing large volumes of air through linseed oil at temperatures ranging from 100° to 140°C.

At one time most letterpress and lithographic inks were based on heat-treated linseed oils and related drying oils. In recent years, printing inks have been required to meet higher standards of performance – both on the faster running presses and also as dry films. These demands for quicker drying high performance inks have led to the development of new vehicle systems, some

166

being chemical modifications of the traditional linseed oil varnish and others completely synthetic media. These are referred to again later in this chapter. Although the chemistry of these materials varies enormously, their basic drying action is similar to that of the natural drying oils.

Non-drying oils

The non-drying oils include the mineral oils, which are high boiling fractions of petroleum. Chemically, mineral oils are mixtures of hydrocarbons and since these contain no reactive groups, they are extremely inert materials. They are available in a range of viscosities, under such names as spindle oil, fuel oil, lubricating oil, and kerosene, and as a group of materials they merge with the hydrocarbon solvents, which are chemically similar but have lower boiling points. Mineral oils are the basic liquid components in news inks and they are also used as solvents and diluents in many inks drying primarily by oxidation.

RESINS

The term 'resin' is not easy to define in a few words. Its meaning is related more to a physical state in which matter exists, than to the matter itself. In other words, the 'resinous' character of these materials is not related to a certain chemical structure. Resins include materials of vegetable and animal origin which are mixtures of several substances, eg rosin, shellac, etc, and also a large number of synthetic polymers of widely differing chemical structure, eg hydrocarbons, alcohols, esters. They range from hard brittle solids to viscous liquids.

In general we can say that resins are organic compounds, of complex structure and high molecular weight, with the ability to dissolve in some organic solvent. When the solvent evaporates from a resin solution so that the resin concentration gradually increases, the resin does not crystallise out, as would an inorganic salt from an aqueous solution, but the solution becomes steadily more viscous until finally, the resin molecules come into close contact with one another and a gel is formed. This property is of importance in many of the applications of resins in surface coatings. A solution of a solid resin in a suitable solvent, eg toluol, methylated spirit, provides the normal vehicle for a gravure, flexographic or any other ink which dries by evaporation. In those inks which dry largely by oxidation, resins may be blended with or chemically combined with natural drying oils in order to improve their performance or alternatively they may provide completely synthetic drying oil systems.

Natural resins

Until early in this century all the resins used in the surface coating industry

167

were of natural origin. A number of these materials are still of importance, although in most cases they are chemically modified before being used in varnishes and inks. Natural resins are generally transparent brittle solids, yellow to brown in colour. Their chemical structure is always complex and in many cases is not accurately known. Most of them are of vegetable origin, being exuded from the bark of various trees. Notable exceptions are shellac, which is the secretion of an insect, and asphalts like gilsonite which occur as minerals.

Rosin, the most important of the natural resins, is obtained from the fluid 'oleoresin', which is tapped from pine trees. When this fluid is heated, turpentine can be distilled off, leaving the pale yellow rosin. Rosin dissolves in alcohol and is of value as a cheap film-forming substance for low cost lacquers. However, its real importance is as a raw material in the production of other resins for printing inks. These include *zinc and calcium resinates* and *ester gum,* which may be used as resins for inexpensive gravure inks, and the rosin-maleic anyhdride adducts (maleic resins) and rosin modified phenol formaldehyde resins much used in all classes of printing inks.

Other natural resins obtained from trees for use in lacquers, varnishes and inks are *Gum dammar, Congo copal* and *Manila copal. Gum arabic,* obtained from certain types of acacia tree, is of interest because it is water-soluble and finds application in lithographic fountain and 'gumming up' solutions.

Synthetic resins

Synthetic resins are polymers made from quite simple chemical substances by the processes of addition or condensation described on pages 42 and 45. Unlike the natural resins, their chemical structure is fairly accurately known and they can be produced to a much closer specification. Furthermore, by selecting different chemical raw materials and varying their relative proportions it is possible to produce resins which are 'tailor-made' for a particular function.

No attempt will be made here to classify the very large number of synthetic resins. Some of these materials were discussed in Chapter 4, including condensation polymers like the polyesters and polyamides, and addition polymers like polystyrene and the polyvinyl resins. Many of the chemical families of synthetic resins are used in printing inks. We will simply consider one such family, the *alkyd resins,* since these form the largest group of synthetic resins available to the surface coating industry and they are consumed in the greatest quantity.

Alkyd resins belong to the larger family of polyesters. We saw on page 46 that long chain polyesters with thermoplastic properties are produced by a condensation reaction between an alcohol with two hydroxyl (OH) groups and an acid with two reactive hydrogens (H). If we represent the alcohol as HO—●—OH and the acid as H—☐—H then the reaction can be shown as

HO—●—OH + H—☐—H ——→ —●—☐—●—☐—●—☐ etc + H₂O

If the alcohol has three OH groups available, like glycerol, then the condensation will produce a polymer network with thermosetting properties (fig. 12.3). This type of *alkyd resin* is of little value in surface coatings because it

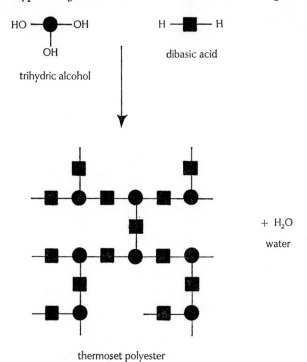

12.3 Formation of a thermosetting polyester.

lacks solubility in the common organic solvents. However this solubility can be obtained by building a long chain fatty acid into the molecule.

If we represent the fatty acid, *eg* linolenic acid, as 〜〜〜H then the modified alkyd can be represented as fig. 12.4

Although the alcohol has three (OH) groups and the acid two (H) groups, the fatty acid only has one reactive group available so its effect is to prevent a three dimensional polymer network building up. These resins are called

169

oil-modified alkyd resins or simply alkyd resins. They are not completely synthetic because the fatty acids used are obtained from natural oils. If unsaturated fatty acids from drying oils are selected, *eg* linolenic acid (page 165), then the alkyd itself will dry by oxidation and polymerisation. Alkyds of this type are widely used in letterpress and lithographic inks. Compared with straight linseed oil varnishes they dry more rapidly and give tougher and more flexible films with superior gloss, partly arising from better pigment wetting properties. Their flexibility makes them particularly suitable for tin printing inks.

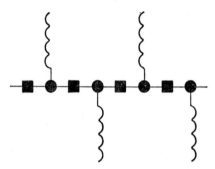

12.4 Structure of oil-modified alkyd resin.

SOLVENTS

Solvents are liquids capable of dissolving other substances, and those used in printing inks are able to dissolve resins or oils. Generally they are organic compounds with much smaller and simpler molecules than the resins which we have just considered. Some of the more important chemical groups of solvents are shown in Table 12.1, with examples of individual solvents from each group.

When a solvent is being selected for a particular ink the chemist has to consider many different factors. First and foremost a solvent's usefulness depends on its *solvent power*, that is, its effectiveness in dissolving a particular material such as an oil, resin or wax. Clearly, it would be very useful if it were possible to accurately predict which solvents would dissolve a given substance, simply from a knowledge of the chemical structure. As in all fields of technology, experience has to be added to theoretical knowledge before the right answers can be given to practical questions of this type. Even then the performance of the solvent or solvents selected must be tested in practice. However, useful guidance on the question 'which solvent?' can be obtained by considering the polarity of the substance which it is required to dissolve.

Table 12.1 Some common printing solvents

Chemical family	Example	Chemical constitution	Boiling point or range	Examples of uses
Hydrocarbons	Petroleum distillates SBP2, SBP3, etc.	Mixture of hydrocarbons Narrow, medium or wide cut fractions from petroleum distillation	e.g.s 71–94°C (SBP2), 100–120°C (SBP3), 240–290°C 280–320°C	Low boiling distillates in flexographic and gravure inks High boiling distillates in letterpress and litho inks
	White spirit	Mixture of hydrocarbons	150–190°C	Silk screen inks Paints
	Toluol (Toluene)	Mainly $\bigcirc$–CH$_3$	105–111°C	Gravure inks
Chlorinated hydrocarbons	Trichlorethylene	$CCl_2 = CH.Cl$	86–87°C	Powerful solvent for oils, fats and waxes. Used as cleaning fluid. Rarely used in formulations
Alcohols	Methylated spirit	Mainly $CH_3. CH_2.OH$ Ethyl alcohol	77–79°C	Flexographic and gravure inks Spirit varnishes
Glycols	Ethylene glycol	$CH_2.OH$ | $CH_2.OH$	197°C	Moisture set and water reducible letterpress inks
Ketones	Acetone	CH_3 $C=O$ CH_3	55–56°C	Flexographic and gravure inks
Esters	Ethyl acetate	$CH_3-C \overset{O}{\underset{O.CH_2.CH_3}{\|\|}}$	76–79°c	Flexographic and gravure inks Lacquers

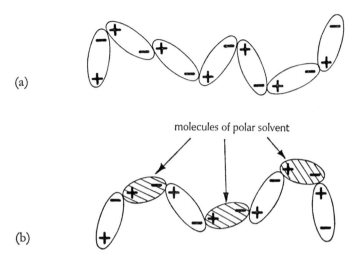

(a)

(b)

molecules of polar solvent

12.5 Attraction of polar molecules.

We saw in Chapter 1 that organic compounds formed by covalent linkages between atoms could have a polar character due to an unequal sharing of electrons throughout the molecule. One effect of this polar character is to increase the force of attraction between neighbouring molecules, the positively charged 'head' of one molecule being drawn to the negatively charged 'tail' of a neighbouring molecule (fig. 12.5a).

Solution can only be achieved when solvent molecules are interspersed between molecules of the substance being dissolved. If the substance has polar character then clearly a solvent with similar polar character has the best chance of breaking up the association between the molecules (fig. 12.5b). In other words, when two substances A and B are brought into contact, solution will only be achieved when molecules of substance A are able to move freely among those of substance B. If the forces of attraction between A and A or between B and B are greater than those between A and B then the two substances will not mix to form a solution.

In practice one does find that substances tend to dissolve best in solvents having a similar polarity. For example, water (polar) will mix with alcohol (polar) and acetone (polar), but not with hydrocarbons like white spirit (non-polar) or benzene (non-polar). Polyvinyl alcohol is one of the few synthetic resins which is sufficiently polar to dissolve in water. Non-polar hydrocarbon solvents like white spirit will dissolve resins like polystyrene (non-polar) and mix with drying oils like linseed oil (weakly polar), but they will not dissolve resins like cellulose acetate (polar) or nitrocellulose (polar). In order to dissolve these polar resins it is necessary to use polar solvents like the esters and ketones. Although there are exceptions, the simple rule that 'like dissolves like' can be a useful guide in the search for a solvent for a particular material.

Solvent power is only one of the factors influencing the choice of a solvent or solvents for an ink. Since the solvent is usually required to leave the ink film immediately after printing, its rate of evaporation is obviously very important. In order to obtain the correct rate of evaporation it is often necessary to use a mixture of solvents. Other solvent properties which have to be considered are their flash point, toxicity, odour and their tendency to leave a trace of solvent trapped in the dry film.

Solutions of resins are often diluted with a second liquid which, though miscible with the solvent, has no solvent power for the resin. Such a liquid is called a *diluent* and it may be added in order to adjust the viscosity, evaporation rate, cost or some other property of the mixture. For example, toluene is often present in nitrocellulose lacquers, although it has no solvent power for the resin. Obviously there must be a limit to the amount of diluent that can be added to the mixture before the resin is thrown out of solution. This type of precipitation is put to good use in moisture set inks. These inks contain a resin dissolved in a water-miscible solvent, *eg* ethylene gycol. Water, acting as a diluent, enters the ink film from the paper or the atmosphere until a point is reached when the resin is precipitated (page 151).

Plasticisers

Plasticisers may be considered as high-boiling, *ie* low-volatility, solvents whose main purpose is to impart flexibility to what would otherwise be a brittle film and to promote adhesion to the substrate. To be effective the molecules of plasticiser must penetrate between the long chain molecules of the resin binder, and in a process similar to lubrication give polymer chains freedom of movement. Only very small losses of the plasticiser by evaporation over a long period can be tolerated or otherwise the film will become brittle. The migration of plasticiser molecules to the surface of a film can also cause embrittlement. Conversely the migration of plasticiser from a substrate into an ink film can cause softening and loss of key of the ink film.

Although most plasticisers are viscous, high boiling liquids, *eg* dibutyl phthalate (b.pt. 340°C) and tritolyl phosphate (b.pt. 420°C), they can also be solids. Plasticisers are incorporated in gravure and flexographic inks and in most surface coatings drying by solvent evaporation.

173

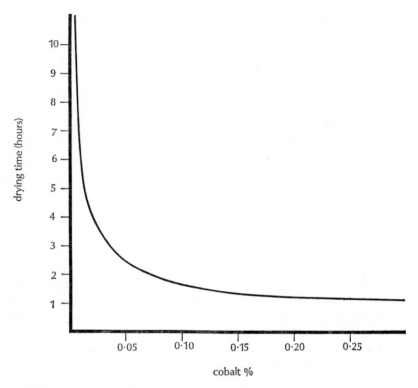

12.6 Drying time on glass of litho varnish containing cobalt linoleate.

DRIERS

The oxidation of a drying oil can be greatly accelerated by the presence of a small proportion of certain salts of a limited number of metals, in particular, cobalt, manganese and lead. The action of these salts or driers, as they are called, is to act as catalysts, that is, substances which assist chemical change without undergoing any change themselves. In this way, driers are able to reduce the drying time of letterpress and litho inks to a few hours.

Driers are used in two different forms, as liquid driers or as paste driers. The more commonly used liquid driers are metallic salts of suitable organic acids, *eg* cobalt linoleate, lead napthenate. These liquid driers are oil-soluble and compatible with drying oil inks. Paste driers are prepared by grinding inorganic salts like lead and manganese borate into linseed oil varnish. They are particularly useful when slow and carefully controlled drying is needed or when a surface receptive to further printing is required, as in much multi-colour printing. In contrast, liquid cobalt driers tend to produce a surface which is not receptive to overprinting. Liquid and paste driers may be combined in the same ink.

The normal range of drier concentration in inks is between 0·5% and 4%. The active portion of this is the metal content, which varies from 4 to 8% in most cobalt driers to 24–32% in lead driers. Higher concentrations in the ink are wasteful since they have little effect (fig. 12.6).

Other additives in printing inks

Small quantities of other materials may be present in printing inks. Natural or synthetic waxes are sometimes included to impart rub-resistance, improved slip and water-repellent properties. Wax compounds may also be added to letterpress and litho inks to reduce their tack without appreciably affecting their flow properties. Antioxidants are organic additives designed to improve the press life of printing inks by delaying the onset of oxidation drying. In order to counter the tendency for a print to 'set off', starch compounds may either be used as spray powders or they may be added to an ink prior to printing. Surfactants, which act in a similar way to soaps and detergents in aqueous solutions, are used as wetting and dispersing agents for pigments in printing ink vehicles.

13. Printing ink
– formulation and manufacture

FORMULATION

The formulation of a successful printing ink is much more difficult than its actual manufacture. In other words, if the ink maker can get his recipe right by selecting the most suitable ingredients in the correct proportions, then the production of that ink in bulk quantities is a relatively simple matter, although variations in manufacturing techniques can cause marked differences in the final result.

In many respects, printing ink formulation is more of an art than a science. Certainly it is a skill based on a thorough knowledge of the physical and chemical properties of materials that can only be developed as a result of a great deal of experiment and experience. No training in organic chemistry, no book on printing inks, and certainly no chapter on formulation can provide a substitute for this experience. The last two chapters have shown something of the range of raw materials available to the printing ink manufacturer. In this chapter we will see how these raw materials are put together in the manufacture of printing inks, and in order to illustrate the problems of formulation we will consider the composition of some typical inks.

Formulation is a complex problem because in selecting from a wide range of materials, the ink maker has so many different factors to take into account. He must select materials which are compatible with one another and which can be satisfactorily blended together into an ink. That ink must first meet the requirements of a particular printing process and type of machine. Then as a dry ink film on a substrate it must stand up to any demands made on it, both during the subsequent conversion of the printed material and during its working life as a printed image. It follows that the ink maker should be given the fullest possible information on how an ink is going to be used. Ideally this should include data on:

(a) the type of press,
(b) the likely printing speed,
(c) the nature of the substrate,

(d) whether the ink forms part of a multicolour printing,

(e) the visual effect required, *eg* gloss,

(f) the nature of conversion processes, following printing, *eg* carton making, varnishing, lamination,

(g) the function of the print, *eg* poster, soap carton, and hence the resistance required to various agencies.

Generally speaking, the more exacting the requirements for an ink, the more expensive it is likely to be, and the overriding factor governing the choice of raw materials is often the price the customer is prepared to pay.

Printing inks can be conveniently divided into two main groups: oil-based inks and liquid inks; although the boundary between these two groups is becoming less precise. The oil-based inks range from the simple news inks containing a large proportion of a non-drying mineral oil, to the letterpress and litho inks based on a drying oil vehicle. On the other hand, liquid inks contain a large proportion of a volatile solvent like toluene or alcohol, and they dry wholly or partly by the evaporation of this solvent. The term 'liquid inks' arises from the fact that these inks are of lower viscosity than the oil-based variety, even the relatively thin rotary letterpress news inks. All the inks used for gravure and flexographic printing are liquid inks.

Oil-based inks

Letterpress and litho inks are basically very similar, but certain differences do arise because of the additional requirements of the lithographic process. For example, the presence of water on a litho press means that inks must have a resistance to bleeding and emulsification, not necessarily shown by a letterpress ink. Since ink film thicknesses are much lower in offset litho than in letterpress, the proportion of pigment may be higher and the particular pigments used must have good colour strength. The viscosity and tack of a litho ink tend to be higher than that of a letterpress ink, partly due to higher pigment concentrations, and partly due to the need to reduce mixing with water.

The compositions of some typical examples of oil-based inks are now given, together with comments on the functions of the various constituents.

1 *Rotary letterpress newsprint black*

Drying mainly by absorption

% by weight	Ingredient	Comments on function, etc
12	Carbon black, news ink grade	Cheap grade of this dense black pigment
3	Induline dye toner in oleic acid	Dye dissolved in a fatty acid which is compatible with ink vehicle, gives blue tone to ink. Stains rollers but this is not important on a continuously black press
80 5	10 poise* mineral oil) 1 poise mineral oil)	Oils of two different viscosities, proportions balanced to give a. correct viscosity of ink to suit press speed b. correct rate of absorption drying

(*Note* If better flow properties are required, 5% asphalt resin, *eg* Gilsonite, may be added with lower 10p : 1p oil ratio.)

2 *Web offset litho newsprint black*

Drying mainly by absorption

% by weight	Ingredient	Comments on function, etc
18	Furnace carbon black	Higher concentration of pigment than letterpress newsprint because of lower film thickness
4	Bronze blue	Blue toner, but pigment rather than a dye, which would bleed into water. Cheap compared to other blue pigments
50 6	10 poise mineral oil 1 poise mineral oil	Proportions balanced to suit press speed and drying rate
12	Hydrocarbon resin	Inexpensive resin which contributes water resistance and pigment binding power. Added as a varnish, dissolved in distillate below
10	Petroleum distillate (fraction boiling 280°–310°C)	Solvent for resin above. Small % added alone to allow fine control of tack and viscosity

* Poise, units of viscosity, where $1 \text{ Ns/m}^2 = 1 \text{ Pas} = 10$ poise, (p.)

3 *Letterpress book-printing black*

Drying mainly by absorption and oxidation/polymerisation

% by weight	Ingredient	Comments on function, etc
18	Furnace carbon black	High % for letterpress in order to give intense black
8	Bronze blue	Toner pigment giving ink a rich blue tone
10	20% asphalt resin (Gilsonite) in 1p mineral oil	Inexpensive resin which assists wetting of carbon black and imparts some extra colour to ink
56	50% esterified rosin-maleic resin in 20p linseed stand oil varnish	Contributes hardness to ink film, also improves wetting and flow properties
2	High boiling petroleum distillate or light mineral oil	For fine control of tack and viscosity
3	Balanced cobalt/ manganese/lead driers	Soluble active driers catalysing oxidation/polymerisation
3	Cobalt paste driers	Feeder driers

4 *High gloss letterpress magenta*

Drying mainly by oxidation/polymerisation

% by weight	Ingredient	Comments on function, etc
12	Calcium 4B toner (resinated)	Better flow properties obtained by using a pigment whose surface has been treated with a resin
5	PTMA Magenta (phospho-molybdo-tungstate complex of a basic dye)	
8	Alumina	Highly transparent extender Controls the balance between colour strength and working properties obtained at a given level of pigmentation

64	Gloss varnish consisting of: 45 parts esterified rosin-modified phenolic resin, 55 parts alkali refined linseed oil. Viscosity of varnish *circa* 200p	The oil and resin are cooked together to give a complex system, partly inter-reacted by ester interchange, which does not show any phase, *ie* oil and resin, separation when on the paper
8	Alkali refined linseed oil	Gives some setting characteristic as well as allowing fine control of viscosity
3	Paste driers	The use of liquid driers tends to be avoided, where wet-on-dry overprinting may follow

5 *Quick-set letterpress blue*

Drying mainly by absorption and oxidation/polymerisation

% by weight	Ingredient	Comments on function, etc
8	Phthalocyanine blue	Two organic pigments from the same chemical family, blended to give the correct hue
2	Phthalocyanine green	
75·5	Quickset varnish consisting of: 40 parts esterified rosin modified phenolic resin, 20 parts isophthalic linseed oil alkyd, cooked together and reduced with	Together providing a rapidly touch dry binder for the pigment, later hardening by oxidation/polymerisation of the alkyd
	40 parts petroleum distillate (fraction boiling 280–310°C) Viscosity of varnish *circa* 200p	This portion rapidly absorbed into paper to give quick set action, ultimately evaporates
10	Alkali refined linseed oil	Assists setting and ultimately dries to assist binding of pigment
4	Petroleum distillate (fraction boiling 280–310°C)	For fine control of viscosity and tack
0·5	6% Cobalt naphthenate liquid driers	Catalyses drying action of alkyd and linseed oil

6 *Web offset heatset blue*

Drying initially by evaporation, with some absorption and final hardening by oxidation/polymerisation

% by weight	Ingredient	Comments on function, etc
12·5	Phthalocyanine blue	Blended to desired hue,
3·5	Phthalocyanine green	heat resistant organic pigments
70	Heatset varnish consisting of:	(cf quickset varnish)
	40 parts esterified rosin modified phenolic resin, heated into 20 parts isophthalic linseed oil alkyd, reduced with	Similar to quickset varnish
	40 parts petroleum distillate (fraction boiling 240–270°C). Viscosity of varnish *circa* 200p	Faster evaporating fraction than in quickset varnish, so faster drying, but also reduced press stability
8	Petroleum distillate (fraction boiling 260–290°C)	For fine control of tack and viscosity, also blended to give correct evaporation rates and press stability for the press/drier conditions involved
5	Petroleum distillate (fraction boiling 280–310°C)	
		Note absence of linseed oil, as in quickset varnish. This would give slow heat drying
1	6% Cobalt naphthenate lipuid driers	Catalyses drying of alkyd, high % for rapid drying

7 Offset metal-printing yellow

Drying mainly by oxidation/polymerisation in a stoving process following printing

% by weight	Ingredient	Comments on function, etc
12	Benzidine yellow	Transparent organic pigment, good colour strength but tends to give poor flow properties
10	Alumina extender	To reduce oiliness and excessive flow of an ink otherwise pigmented at a low level, improves ink distribution and slightly reduces price
26	Medium-long oil alkyd (*circa* 300p)	Rapid drying to a hard film
24	Maleic resin-alkyd varnish, *eg* 45 parts esterified rosin-maleic resin, dissolved in	} Contributes hardness
	45 parts dehydrated castor oil-pentaerythritol-phthalic anhydride alkyd, *circa* 150p.	Alkyd gives toughness and flexibility
	10 parts petroleum distillate (fraction boiling 280–310°C)	Thins varnish to about 200p. Slow evaporating at normal temperatures but evaporates during stoving
22	Dehydrated castor oil-pentaerythritol-phthalic anhydride alkyd modified with tung oil, *circa* 200p	Combined with other alkyds to give the required rate of drying and degree of toughness
5	Petroleum distillate (fraction boiling 280–310°C)	For fine control of tack and viscosity. Evaporates during stoving
0·4	6% manganese naphthenate driers	Low concentration, but sufficient to bring about drying during stoving
0·6	Antioxidant medium, *eg* oil of cloves (10%) dissolved in distillate (90%)	Prevents ink drying on the press, but evaporates during first few seconds of stoving

Liquid inks

Liquid inks are basically pigment/resin/solvent mixtures. Because of the wide variety of substrates to which they are applied, a large number of different resin binders and hence solvents are used. In formulating liquid inks, particularly for packaging applications, care must be taken to avoid the problems of solvent retention in the ink film and of residual odour. Although gravure and flexographic inks are of similar formulation there are certain points of difference. For example, flexographic ink films are often thinner than those for gravure, so the colour strength of flexographic inks may need to be higher. The solvents used in flexographic inks must not attack rubber stereos and this rules out solvents like toluene. Both gravure and flexographic inks are usually supplied at higher than press viscosities, and then reduced with solvent before printing.

The composition of typical examples of gravure and flexographic inks now follow.

8 *High quality gravure ink for paper*

% by weight	Ingredient	Comments on function, etc
11	Furnace carbon black	High quality grade with good colour strength and flow properties, but less expensive than channel black
4	PTMA blue (Phospho-molybdo-tungstate complex of a basic dye)	Expensive, but low % sufficient to give blue tone to ink
20	Chlorinated rubber resin	Excellent tough film former
10	Esterified rosin-maleic resin	Reduces resin cost slightly and improves gloss
2	Polyethylene wax	Included when good rub-resistance is required
53	Toluene	Solvent for above resins. Grade used which is free of sulphur, benzene and residual odour

9 *Flexographic red for paper*

% by weight	Ingredient	Comments on function, etc
15	Permanent red 2B	Red organic pigment
10	Titanium dioxide	Opaque white pigment reducing red shade
10	Nitrocellulose	Imparts good wetting and flow properties
13	Shellac	Provides tough film, with good key to substrate
2	Dibutyl phthalate	Plasticiser for nitrocellulose. Higher proportion required for inks printed on to plastic films
27	Methylated spirit	Good solvent for shellac but only diluent for nitrocellulose
14	Ethyl acetate	Good solvent for nitrocellulose, but only tolerated by shellac. Fast evaporating solvent
5	'Cellosolve' (ethylene glycol mono-ethyl ether)	Solvent for both nitrocellulose and shellac. Used to adjust drying rate of ink, fairly slow evaporating solvent
4	Wax compound	A medium of polyethylene wax in a suitable vehicle, used to give surface slip and rub-resistance to dry film

(*Note* With a slight upward adjustment of plasticiser this ink would be suitable for printing cellulose film, *eg* moisture-proofed 'Cellophane'.)

MANUFACTURE

The basic objective in printing ink manufacture is to produce an efficient dispersion of pigment particles in a varnish. The machinery used is relatively simple, but the organisation of this plant is more difficult, because in providing a service to the printer, the ink maker has to be ready to supply a large number of different inks, often in relatively small quantities and at short notice. In providing a service to the printing industry, many ink makers also manufacture and supply such things as rollers, litho plates, set-off sprays, etc, in addition to printing inks and other surface coatings.

Letterpress and litho inks

The manufacture of letterpress and litho inks usually takes place in two stages – mixing and milling. Mixing is a process in which the constituents of the ink are mechanically blended so that the pigment is initially wetted, air in the mixture displaced and the aggregates of pigment reduced in size. After mixing, the ink should be reasonably uniform in colour and consistency, but it may still contain relatively large particles. The subsequent process of milling brings the ink to a smooth and homogeneous state, ready for use on a printing machine. In recent years the trend in ink manufacture has been to increase the dispersion efficiency of mixing machinery and so allow a reduction in the subsequent milling time.

Mixing The mixing of pigment into a letterpress or litho varnish was originally carried out by hand stirring or by the direct addition of the materials on to the back pair of rolls of a three-roll mill (fig. 13.2). Mixing on an open mill has a number of disadvantages. Some pigment is bound to be lost in the form of a fine dust, and the mill needs resetting when mixing is completed and dispersion commences. These problems are overcome by introducing a separate mixing operation, in which pastes are subjected to shearing stresses applied by strong blades rotating vertically or horizontally in heavy-duty plant.

Mixers in which blades rotate about a vertical axis are used for the softer pastes. In the so-called change-pan mixers the shaft is driven from above so that the stirrer can be lifted or the pan lowered (fig. 13.1). In order to ensure that there is a vertical as well as a rotary movement of the paste, the blades are set at an angle. In recent years the construction of mixers of this type has been strengthened and they are now able to handle all but the stiffest pastes.

For mixing the stouter pastes it may be necessary to employ a heavy duty Z-arm dough mixer. These normally have two Z-shaped blades rotating horizontally at different speeds and in opposite directions, in a container which is shaped so that the blades scrape its bottom surface.

Milling Milling has been defined as 'dispersing pigments in vehicles to a point at which the finished ink gives satisfactory results on the printing machine for which it is prepared'. As this definition implies, the degree of dispersion required is not the same for every type of ink, and in fact complete dispersion is neither attainable nor desirable. There are many different types of ink mills, but in every case dispersion is achieved by a combination of shearing and crushing forces.

Paste milling for letterpress and offset litho inks is usually carried out

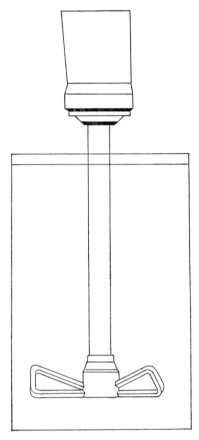

13.1 Heavy duty paste mixer (Torrance).

on roll mills. These are machines, consisting of a number of metal rolls driven in contact with one another at different speeds. The ink paste is fed into a nip between two of the rollers, and leaves the mill via a scraper knife, resting on one of the rollers. The three-roll mill is the standard piece of equipment for milling letterpress and litho inks. At its simplest, it consists of three steel cylinders of identical size mounted parallel to one another (fig. 13.2a). The rolls are driven at different speeds, which increase from the back to the front of the machine. Because heat is generated by the milling action, the rolls are made hollow so that they can be continuously cooled by water. The two back rolls are fitted with a hopper into which the paste is fed. The paste is carried forward into the nip between the front two rollers and finally is

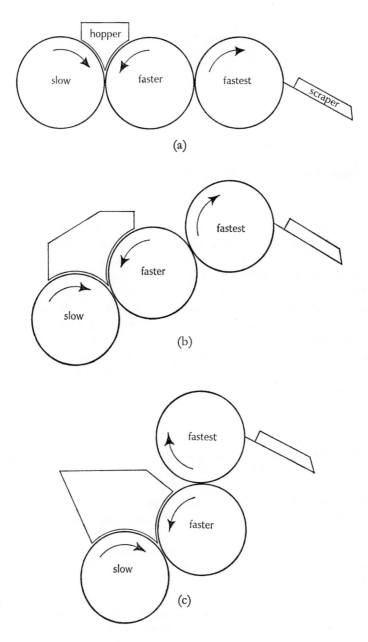

13.2 Various arrangements of three roll mills.

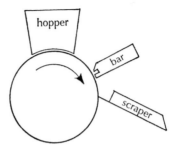

13.3 Single roll mill.

removed from the mill by the scraper blade set at an angle to the front roll. The distances between the rollers and hence the degree of milling can be controlled by moving the rolls relative to one another. The horizontal in-line system of rollers shown in fig. 13.2a has a number of disadvantages, and other arrangements of the three rolls are more popular. With the back rolls in the positions shown in figs. 13.2b and 13.2c, the operator has a shorter distance to lift the ink in feeding the mill, feeding can be less frequent because the staggered arrangement of the rolls provides a nip which is capable of holding a larger quantity of ink and the mills occupy less floor space. Single roll mills, in which the nip is between a rotating roll and a stationary bar are suitable for low viscosity mixtures including news ink (fig. 13.3). Other mills with two, four or five rolls are also used to a limited extent.

Liquid inks

The methods of manufacturing liquid inks for gravure and flexography differ from those already outlined for letterpress and lithographic inks for two main reasons. Firstly, liquid inks contain appreciable quantities of volatile solvents, so their exposure to the atmosphere must be kept to a minimum during manufacture in order to avoid solvent loss. Secondly, they have a much lower viscosity so that high speed mixing and milling equipment can be used.

The stages of mixing and milling liquid inks are often combined in a single operation. Where premixing is necessary, use is made of high speed vertical mixers. The impeller may be driven from above, like the change-pan paste mixers already described, or they may be driven from below (fig. 13.4). In this latter type, the mixture has to be discharged from an outlet pipe, with or without the assistance of a pump. Normally the use of these high-speed vertical mixers would be followed by some form of milling. They are also used to

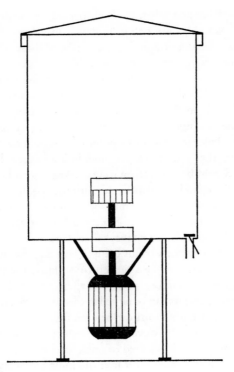

13.4 High-speed mixer.

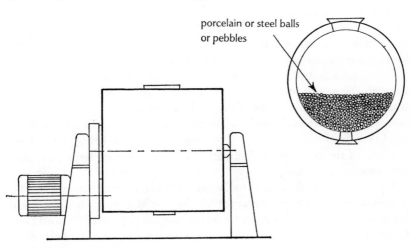

porcelain or steel balls
or pebbles

13.5 Principle of the ball mill.

manufacture varnishes and when easily dispersed pigments are involved or when pigment/resin chips are being dissolved in solvents, it is possible to complete the manufacture of a liquid ink in one of these mixers.

Ball mills, which combine mixing and milling processes, have provided the traditional method of dispersing pigments, resins and solvents in the manufacture of liquid inks. Basically they consist of steel cylinders which are partly filled with steel or porcelain balls or pebbles together with the ingredients of the ink (fig.13.5). On rotating the cylinder about a horizontal axis, the grinding balls are raised to a point from which they come cascading down over one another and against the inner lining of the cylinder, subjecting any ingredients caught between them to both shearing and crushing stresses. The rotation of the cylinder is continued until the required degree of dispersion is obtained, a period which may be as long as 24 hours. Variables influencing the efficiency of ball milling include the rotational speed of the cylinder, the ball size and shape, the ball charge volume and weight, the nature and proportion of the various components and the temperature of the charge, and these factors can be adjusted to optimise the performance of a mill. Ball mills have a number of advantages: no preliminary mixing is necessary, all the ingredients can be added at the same time, the mills are easily loaded and discharged, night hours can be used for production and labour and maintenance costs are low. However, on the debit side, the rate of production is very slow, the method is less suited to small batches, only about 25% of the cylinder's volume can be filled with ink, the mills are noisy and difficult to clean and inks can become contaminated with abrasive particles from the balls or from the cylinder lining. These disadvantages have led to attempts to incorporate some of the principles of the ball mill into more productive plant for milling liquid inks and the successful development of such equipment has led to the replacement of the ball mill for many applications.

One approach to the problem of speeding up the milling action of ball mills was to use a stationary vessel and drive steel or ceramic balls rapidly round the container by means of a central spindle. This is the principle of the *attritors*, which are available in a range of sizes, and also in a form which allows continuous operation, with pre-mixed ingredients being pumped into the base of the container and dispersed ink led out from the top, suitable filters retaining the balls.

Sand or bead mills work on a similar principle to the continuous attritor, but instead of using the relatively large balls they use sand or beads as the grinding medium. The sand is driven round the container by plates attached to the central rotating cylinder and dispersion takes place as a result of the differences in speed between the moving layers of sand. Continuous attritors and

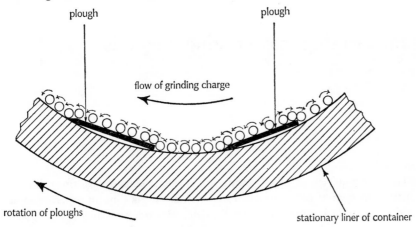

plough   plough

flow of grinding charge

rotation of ploughs   stationary liner of container

13.6 Principle of the microflow mill (Torrance).

sand mills can be used for news ink manufacture and steel shot loaded versions can cope with low viscosity forms of web offset ink.

The development of *microflow mills* has provided another method for the continuous dispersion of liquid inks. Their unique feature is a rotor carrying a series of ploughs round the periphery of a hollow horizontal cylinder. The leading edge of each plough is tapered to act as a scraper, and in place of balls or sand, stainless steel rods act as the grinding medium. As the rotor turns, the rods are thrown to the outside of the rotor where they form a single layer against the ploughs and the inside of the cylinder (fig. 13.6). The difference in speed between the ploughs and the rods, and the change in the direction of rotation of the rods as they pass over the ploughs causes intense shear in the thin layer of ink in which they are turning. The rate at which ink flows out of the mill is adjusted according to the degree of dispersion required. The first microflow mill was introduced for small batch production but it is now available in a larger size, which can be used for the continuous or batch production of liquid inks. The premixing of pigment and vehicle is required.

Increasing quantities of liquid ink are now manufactured from 'chips', produced by directly dispersing pigment into solid resin. The dispersion is carried out on a heavy two roll mill, similar to those used in milling rubber, the resin becoming fluid under the high temperatures and pressures involved. The powerful milling forces ensure the maximum breakdown of pigment aggregates and simplifies the subsequent manufacture of liquid ink from the chips. This is carried out on a high speed mixer which disperses the chips in the solvent and the additional varnish required in excess of that given by the resin in the chip. Inks produced from chips are said to have improved gloss and colour strength and cause less wear on gravure cylinders.

14. Printing ink – testing methods

The first requirement of a printing ink is that it should perform well on a particular type of printing machine. Later, as an ink film on paper or some other substrate, it must stand up to the demands made on it during subsequent processing and during its use as a print. Methods of testing inks have been developed in an attempt to predict the performance of an ink during these various stages of its working life, and to provide a means of checking whether an ink has been produced and supplied to an agreed standard.

Testing methods for printing ink can be conveniently considered in two groups according to whether they are carried out (a) on the printing ink itself, or (b) on a print, made with the printing ink.

TESTS ON PRINTING INK

Ultimately the only way to test the press performance of a printing ink is to print with it on the type of machine which is later going to be used. Earlier in this book a similar statement was made about the testing of paper. Both are statements of the obvious, but their truth has dawned comparatively recently. Certainly today the value of the printing machine as a 'test instrument' is acknowledged by the printing ink manufacturers and they have some impressive pressrooms to prove it. Obviously it would not make economic sense for an ink manufacturer to test every batch of ink on a printing machine. Even if it were possible, the results of the tests would only be of value if the printing conditions and the method of assessing the print had been standardised. The large number of variables in the printing operation on a commercial machine make it extremely difficult to establish these standard conditions. The number of variables can be reduced by using the simplest type of printing machine – a proof press, or alternatively by using one of the small printing devices, like the I G T printability tester, which allows the close control of the main printing variables. In this way printing tests can be carried out in the laboratory under standard conditions, but unfortunately it does not follow that the results obtained will simply scale up to allow the accurate prediction of the ink's performance on larger printing machines.

The economic and technical problems associated with printing tests underline the value of identifying those specific properties of an ink which appear to be important on a press, and of developing simple tests to measure these properties. For example, the tack of a letterpress or litho ink would seem to be one such property, and several testing methods have been developed to measure tack. Although a tack measurement only provides a small fraction of the total information about an ink, it is undoubtedly relevant to press performance and at the very least it is a useful check on whether a batch of ink has been made to an agreed standard.

In discussing paper testing earlier in this book, the importance of a proper sampling procedure was stressed. It is equally important that tests on printing inks are carried out on samples that are truly representative of the batch of ink supplied. For example, it may be necessary to take samples from several different tins and from different positions in these tins. Ink taken from near the top of a container may contain more air or more particles of skin than the bulk of the ink. Liquid inks should be thoroughly stirred before sampling, and the sample properly stored in a sealed container until the tests are carried out.

Before considering the measurement of properties which appear to relate to printing performance, we should first consider some of the simple and largely subjective tests which are primarily used by ink makers as a means of checking the progress of manufacture. As these tests may be carried out before the ink is taken off the mill so that adjustments can be made if necessary, they must be carried out quickly.

SUBJECTIVE METHODS OF TESTING

As in the case of paper, testing methods for ink range from simple subjective tests using little more than the five senses, to sophisticated methods which rely on expensive instruments. We saw in an earlier chapter that an experienced paper technologist can find out a great deal about a piece of paper, simply by using his hands, eyes and occasionally ears. Much the same is true for a printing ink technologist providing he is armed with a palette knife, a draw-down knife, some paper and a sample of ink. His tests would normally be carried out on a standard ink as well as on the test sample, so that the properties of the two inks would be directly compared. As far as the 'flow properties' of the ink are concerned, it is possible to compare the resistance offered to the motion of a palette knife, the length of the string drawn out when the knife is rapidly pulled away from the ink, and the rate at which the ink flows off the knife. It is important to remember that before any comparative test is made, the two inks should be vigorously worked with the knife, in order to break down any internal structure which can be set up in an ink when it is left undisturbed. A rough assessment of an ink's *tack* can be made by using a finger to dab out films of the two inks side by side. If these dabbed-out films are of approximately printing thickness, they can also be used to gauge the *relative hue and colour strength* of the ink. However, this comparison is usually made by pulling a 'drawdown' of the two inks. This is most simply done by placing a small amount of each ink about an inch apart at the head of the top sheet of a firmly clamped pad of paper, and then using a draw-down knife to spread the two inks over the paper, in such a way that the two ink films have a common boundary. The knife is drawn firmly towards the operator with the blade in an almost vertical position. Towards the end of the stroke the pressure can be relaxed and the angle of the blade lowered so that a thicker film of the two inks is obtained at the bottom of the sheet of paper. It is important that the inks being compared are of similar consistency, otherwise the thicknesses of the two films will be different.

Ideally drawdowns should be made on pads of the paper on which the ink is likely to be printed, but in practice pads of a reasonably smooth paper, *eg* a super-calendered grade, are often used. Draw-down pads with a black band of ink printed across the middle of the sheet are useful when the opacity of two inks is being compared.

Some quantitative assessment of the relative colour strength of two inks can be obtained by reducing both inks with an equal quantity of an opaque white ink (*eg* 10 parts of zinc oxide in stand oil to 1 part of coloured ink).

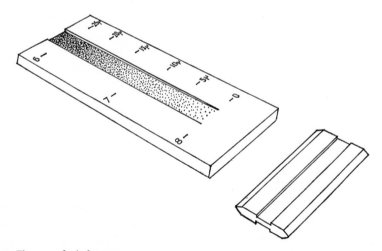

14.1 Fineness of grind gauge.

The two tints are then viewed side by side on a glass slide. If they only differ in strength, then it is possible to add measured volumes of white ink to the stronger sample until a match is obtained and the relative colour strengths can be calculated.

The value of these simple techniques for comparing the consistency, hue and colour strength of an ink with a known standard, depends to a large extent on the experience of the person carrying them out. Many other ink tests rely on the judgement of the human observer, although not to the same extent. Although the measurement of the *degree of dispersion* of the pigment into the vehicle during the milling of a letterpress or litho ink makes use of an instrument, considerable experience is necessary in interpreting the results. These measurements are usually made on a 'fineness of grind gauge', a small block of hardened steel which has machined into it a wedge-shaped channel about 200mm long varying in depth from zero at one end to 25μm at the other (fig. 14.1). A small amount of ink is placed in the deepest part of the groove and a steel scraper used to draw the ink down the gauge so that it completely fills the groove. Any large pigment aggregate left in the ink will be drawn forward by the scraper leaving a scratch in the surface of the ink film. The degree of dispersion can be assessed by observing the distance along the gauge at which the scratches first occur.

OBJECTIVE METHODS OF TESTING

The trend in recent years has been to the development of more objective instrumental methods of measuring specific properties of ink which appear to be relevant to its performance on the press. As was stated earlier, although the information obtained from these tests may not take us to the point where we can predict this press performance, they collectively provide a means of checking whether a sample of ink measures up to an agreed standard. The tests that we shall be considering in this group cover the following properties:

Flow properties: Viscosity; yield value, plastic viscosity, thixotropy. Tack.
Press stability
Drying time on paper
Lithographic performance
Printability
Ink film thickness

Flow properties

The increasing importance attached to the flow properties of fluids in a wide range of industries, with products as different as bread, lubricating oil, chocolate and cheese, has led to the rapid growth of a branch of physics known as rheology, a word meaning 'the science of flow'. In the printing industry the flow properties of inks, varnishes, adhesives and plate coatings have an important bearing on their performance, and the increase in printing speeds and in the use of multi-colour machines has emphasised the need for inks to be supplied with rheological properties falling within fairly narrow limits. Before we consider the testing methods that are used in the control of these properties, we must first consider what these rheological properties are in the case of printing inks.

When a simple liquid like water or glycerine is stirred, the rate at which it flows is directly proportional to the force applied.

Force applied (shear stress) = constant × rate of flow (rate of shear)

$$\frac{\text{Shear stress}}{\text{Rate of shear}} = \text{constant}$$

This constant is known as the *viscosity* of the liquid and its units are Ns/m^2 (formerly poise, p, where $1Ns/m^2 = 10p$). Some idea of the value of these

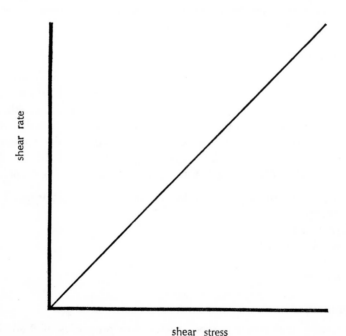

shear stress

14.2 Stress/strain diagram for a Newtonian liquid.

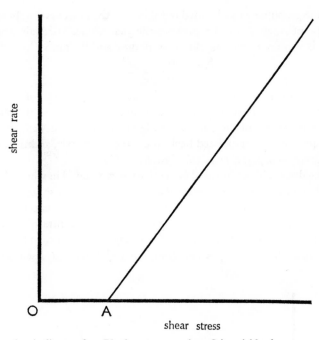

14.3 Stress/strain diagram for a Bingham system, where OA = yield value.

units can be gained by reference to the viscosity of some simple liquids, *eg* water at room temperature, 0·001 Ns/m² (0·01p); high speed gravure ink, 0·01–0·03 Ns/m² (0·1–0·3p); glycerine, 1·5 Ns/m² (15p).

The direct relationship between shear stress and rate of shear for these simple liquids is shown graphically in fig. 14.2. Because the viscosity is the same at all rates of shear the graph is a straight line. The fact that the line passes through O indicates that the liquid will flow, however small the force applied. Liquids which behave in this manner are known as Newtonian liquids and are said to exhibit Newtonian flow.

With the exception of gravure and flexographic inks which are approximately Newtonian in their behaviour, most of the inks, varnishes and adhesives used in the printing industry behave in a rather different way. When pigment is added to a Newtonian liquid, the solid particles increase the resistance to flow and the viscosity increases. Most letterpress and litho inks contain a large volume of pigment and they will not flow at all until the force applied is greater than a certain minimum, which is called the *yield value*. When the shear stress is plotted against the rate of shear as in fig. 14.3, a straight line graph may be obtained but this time the line cuts the shear stress axis at a point A, so that OA represents the yield value. Materials which behave in this way are known as 'Bingham Bodies' and the ratio

$$\frac{\text{Shear stress} - \text{OA}}{\text{Shear rate}}$$

is sometimes incorrectly called the *plastic viscosity*. A third type of flow, quite frequently found in letterpress and litho inks, is one where the viscosity diminishes as the rate of shear increases (fig. 14.4). The straight line portion of the graph can be extrapolated back to cut the stress axis, and the slope of this line gives *an apparent or plastic viscosity*.

Occasionally, highly pigmented systems are encountered in which the resistance to flow increases with increasing shear stress (fig. 14.5), *ie*, the harder you stir the liquid, the more difficult stirring becomes. This effect, known as *dilatancy* is comparatively rare in printing ink systems. In contrast, many letterpress and litho inks have the property of building up resistance to flow, which is broken down by shearing or agitation, but which rebuilds itself when the

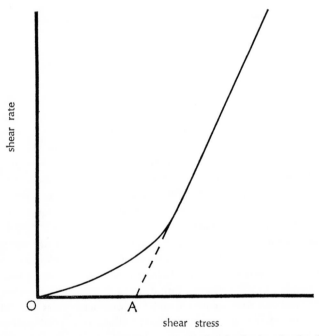

shear rate

O   A

shear stress

14.4 Stress/strain diagram for an ink, whose plastic viscosity is given by the slope of the straight line portion.

198

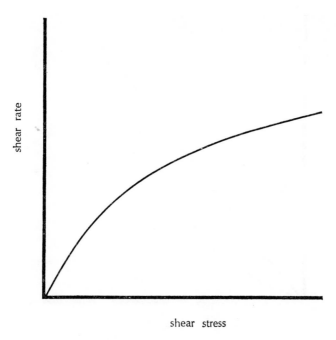

14.5 Stress/strain diagram for a dilatant system.

ink is allowed to stand. This effect known as *thixotropy* has been successfully exploited by paint manufacturers in the familiar non-drip paints. The graph shown in fig. 14.6 is produced by first gradually increasing the rate at which the fluid is sheared and then slowly reducing the rate of shear. For any given rate of shear there are two values obtained for the viscosity – one on the upward slope of the curve and a lower one on the downward slope. If the fluid is allowed to stand for a period of time and the measurements are repeated the same flow pattern is produced. The time taken by the fluid to return to its original high viscosity may vary from a few seconds to several days.

The ability of an ink to flow is obviously essential to its effective performance as it moves from the duct via the rollers and the printing surface to the paper. In making this journey a film of ink is also required to *split* each time it passes from one roller to the next, and at the stages of transfer between the inking roller and the plate, the plate and the blanket in offset printing and finally in the transfer to the paper. The tackiness or *tack* of an ink may be defined as the resistance to splitting offered by an ink during these stages of the printing process. Tack and viscosity are quite distinct properties of an ink.

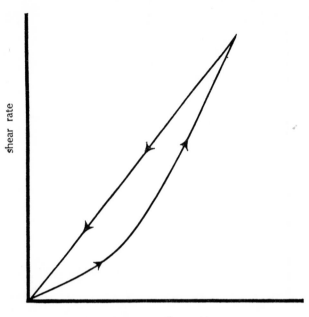

14.6 Stress/strain diagram for a thixotropic system.

If a particular ink is reduced, there is a fall in viscosity and a corresponding fall in tack, but when different ink systems are compared there does not appear to be any clear relationship between the two properties. The tack of an ink is important for a number of reasons. In the moment following printing impression, splitting should occur within the ink film, but if the tack of the ink is high in relation to the surface strength of the paper, splitting may take place in the paper itself and picking occurs (page 130). One other possibility is for the ink to fail to wet the surface of the substrate so that splitting occurs at the boundary between the ink film and the substrate, giving non-trapping (fig. 14.7). The tack value of an ink takes on an even greater significance in wet-

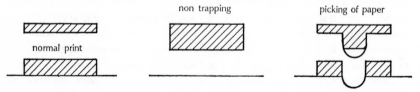

14.7 Non-trapping and picking.

on-wet multi-colour printing. Here it is essential that in the printing of the second and subsequent colours splitting occurs in the film of ink being printed rather than within the film of the previous colour applied or between the two wet films. In order to achieve satisfactory trapping, inks are normally tack graded so that the first ink down has the highest tack and the last ink down the lowest tack.

Since gravure and flexographic inks are virtually Newtonian in their behaviour and have a resistance to flow which is largely independent of the rate of shear, their viscosity can be measured in simple apparatus at a single rate of shear, which need not be accurately known. This is not the case for letterpress and litho inks which behave as non-Newtonian fluids, exhibiting pseudoplastic and thixotropic flow. Here it is necesssary to measure viscosity at several rates of shear and collect data to establish the yield value and apparent viscosity of the system. A number of instruments are now available for measuring the tack value of a letterpress or litho ink. These will be referred to after we have considered methods of measuring viscosity.

The instruments used for measuring the viscosities of fluids are known as viscometers. These fall into three main groups:

(a) Capillary viscometers.
(b) Falling body viscometers.
(c) Rotational viscometers.

Capillary viscometers These depend upon the fact that the time taken for a fixed volume of fluid to flow through a narrow tube varies with its viscosity. The *U-tube viscometer* (fig. 14.8) is a simple instrument of this type, widely used in many industries to measure or compare the viscosities of liquids. The viscometer is filled to a marked level L, then this fixed volume of fluid is drawn up into the bulb T and the fluid is allowed to flow under gravity through the narrow tube VX. The time is taken for the level of the fluid to fall from the mark U to the mark V, and the viscosity of the fluid obtained by multiplying this time in seconds by a factor for the instrument. By choosing U-tubes of appropriate dimensions it is possible to make accurate measurements of the viscosities of oils and varnishes, but in general these instruments are not suitable for printing inks or adhesives. *Efflux viscometers* also depend on the measurement of the time required for the passage of a fixed volume of fluid through a small opening. They consist essentially of containers having a circular opening of standard size in their base. Unfortunately there are many different types of efflux cups with a variety of dimensions giving different values of efflux time for a given viscosity. The *Ford* series of cups (fig. 14.9)

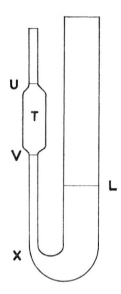

14.8 U-tube viscometer.

are used extensively in the paint industry, other varieties including Redwood, Saybolt and Engler. The *Zahn* cups (fig. 14.10) are probably those most often encountered in the printing industry. Ranges of Ford and Zahn cups are available, numbered according to the size of the hole drilled into the base. Viscosity is measured by filling a cup and then taking the time for the steady stream of fluid coming out of the base to break into a series of drops. The time

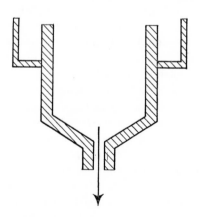

14.9 The Ford cup.

in seconds coupled with the type and number of the instrument is quoted as the measured viscosity. Efflux viscometers are not noted for their accuracy but they are widely used for control purposes on account of their cheapness, simplicity and convenience. As far as printing inks are concerned their use is limited to gravure and flexographic inks with properties very close to those of Newtonian fluids and they are not suitable for the stiffer inks used in letterpress and lithography.

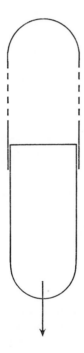

14.10 The Zahn cup.

Falling-body viscometers These rely upon the fact that when a body falls freely through a fluid medium, its velocity depends partly on the viscosity of the fluid. In practice, measurements of the time taken for a heavy body (usually a sphere or rod) to fall a fixed distance through a sample of the fluid are used to calculate the viscosity. In the *falling sphere viscometer* the fall of a stainless steel ball, through a sample of fluid in a glass tube, is timed between two marks engraved on the tube. This type of instrument can be used to make fairly accurate measurements of the viscosities of oils and varnishes, providing these are reasonably transparent. Another instrument in this group,

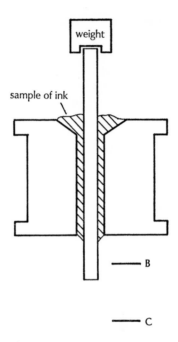

sample of ink

weight

B

C

14.11 Principle of the Laray viscometer.

the *falling rod viscometer* is particularly useful for measuring the flow properties of letterpress and litho inks. The *Laray viscometer* consists of an accurately machined rod which falls through a metal collar, whose internal diameter is only slightly greater than that of the rod (fig. 14.11). The rod is coated with the ink to be tested, any excess ink being collected in the small recess at the top of the collar. The time taken for the bottom of the rod to pass between the marks B and C is measured with a stop watch or in some more sophisticated instruments by an electrical timing mechanism. The rate of shear can be varied by loading the top of the rod with additional weights, so that the instrument may be used with non-Newtonian inks. From the data collected, the apparent viscosity and yield value of an ink can be obtained by reference to a chart. The Laray viscometer is becoming increasingly used for control purposes in the printing industry. It is robust, simple to operate, easy to clean and requires a smaller sample of ink than most viscometers. One limitation of the instrument is that it cannot be used to measure the viscosity of an ink after known rates of shear have been applied for chosen periods, so it does not allow a study of thixotropic behaviour.

Rotational viscometers

In the third group of *rotational viscometers*, the most usual form consists of two cylinders, an outer cylindrical cup which is rotated by a motor and an inner cylindrical bob suspended by a torsion wire or spring (fig. 14.12). The rotation of the outer cylinder sets up a drag on the inner cylinder, which depends on the viscosity of the sample between the two cylinders. By multiplying the consequent deflection of the scale attached to the wire, by a factor for the instrument, the viscosity at that rate of shear can be obtained. By varying the speed of the motor and the relative sizes of the two cylinders this type of viscometer can be used to investigate fluids with widely varying viscosities, over a range of shear rates. Thixotropic flow is shown up by the deflection dropping back as the time of shear increases, and the time taken for the viscosity to return to its original value can be found by allowing varying periods of rest between measurements. Portable versions of this type of instrument can be used to measure the flow properties of adhesives in the

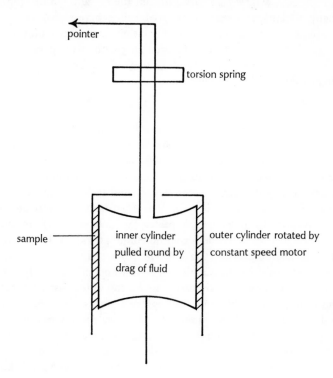

14.12 Principle of a rotational viscometer.

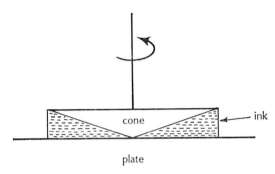

plate

14.13 Principle of a cone and plate viscometer.

tank of a glueing machine or of liquid inks in the duct of a printing machine. Many different versions of rotational viscometers have been developed. In some of these the inner cylinder is rotated and the viscous drag on the outer cup is measured. *Cone-and-plate viscometers* like the Ferranti–Shirley (fig. 14.13) have many advantages over the co-axial cylinder versions, but they are normally used as instruments for research rather than control purposes.

The viscosity of any fluid varies considerably with its temperature and it is therefore essential that all the methods of measuring viscosity are carried out under standard conditions of temperature.

Tack

We have already seen that the tack of an ink can be roughly assessed by an experienced operator by means of a simple finger test in which the ink is dabbed out on a glass plate or slab. The value of this simple test is considerably increased if the tack of the ink is compared with that of a sample ink of known tack. Unfortunately, the consistency and tack of a normal printing ink is liable to change during long periods of storage, and to overcome this problem, specially formulated tack pastes (*eg* Mander-Kidd) are available, which are designed to be stable for an indefinite period. A device known as the NPIRI–Chatillon Tack-Finger (plate 16) was introduced in order to bridge the gap between the simple finger test and the relatively expensive tackmeters. A small recessed circular plate on top of the instrument is inked up using a drawdown technique, the operator places his finger on the ink surface and then raises it as rapidly as possible. After repeating the film splitting process a specified number of times the maximum force is read directly on the dial gauge and this is taken as the tack of the sample.

A roller instrument for the measurement of tack, the GATF Inkometer, was introduced as long ago as 1938. More recently, three other tackmeters have become available – the PIRA tackmeter, the MB Churchill tackmeter and the Tackoscope. In all these instruments the tack value, the force required to split the ink film, is given by a measurement of the drag exerted by a motor driven roller on a rider roller (fig. 14.14). The more recent instruments use more sophisticated methods of measuring this drag. The driven roller is made of metal and water-cooled internally to ensure that the ink film is kept at a constant temperature. Its speed of rotation may be varied so that tack values can be measured over the range of velocities encountered on printing machines. The rider roller may either be made of metal or synthetic rubber. More detailed information on these tackmeters can be found by reference to the sources listed on page 257. Instruments of this type are finding an increasing use with both ink makers and printers in the control of the flow properties of printing inks.

Press stability

The stability of a letterpress or litho ink on the rollers of a press during a production run can be predicted by measuring the change in its tack during an extended run on an inkometer or tackmeter. If the tack is plotted against the time of the run, an ink with good press stability should give a long flat curve. A curve rising rapidly to a high peak may indicate that the tack of the ink would be unacceptably high at impression.

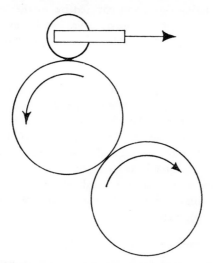

14.14 Principle of tackmeters. The two large inked rollers tend to pull the small roller forward. The force applied to prevent movement is measured to give the tack stress of the ink film.

Instruments, which measure the drying time of a film of ink on glass, have been in existence for some considerable time. In the PIRA Ink Drying Time Tester (plate 17) an ink is rolled out on to a circular glass plate P, which is mounted on a turntable driven by a motor. As the disc rotates a stylus S draws a spiral track on the wet film of an ink. As long as the film is wet, the track has smooth edges, but a point is reached when the ink is sufficiently dry for the film to be torn by the stylus, leaving a track with jagged edges. Subsequent inspection allows the time at which this change takes place to be read off and this is taken as the drying time of the ink. The BK Drying Time Recorder (plate 18) is another instrument of a similar type. A chromium plate carrier draws six hemispherical needles along six rest strips, so recording the drying characteristics of the coatings against a time scale, which runs along the length of the instrument. The graduated knob on the front of the recorder allows a choice of three different speeds of travel.

Drying time on paper

Although the drying time on glass can give some useful information on the press stability of an ink, it has very little bearing on the drying time of the ink on the paper on which it will be printed. A number of separate instruments have been developed for this purpose. In the PIRA Print Drying-time Instrument (plate 20), a long strip of each print is mounted parallel to the axis of a slowly rotating drum and covered by another strip of unprinted paper. As the drum rotates at the rate of one revolution every nine minutes, a weighted stylus is moving along the drum at a rate of 3·8mm/hr so that every nine minutes the unprinted paper is pressed against another small portion of the drying print. As long as the print is wet, a set-off mark appears on the unprinted paper. These marks join up to form a solid band which only ends when the print is dry enough not to set-off, and the length of the band is a measure of the drying time of the ink on the paper. It is important that the test strips are printed at a known film weight. The IGT printability tester may be used for this purpose, and it is also possible to use this instrument to measure the print drying time by carrying out a series of set-off tests against unprinted strips of paper at suitable intervals of time. The final hardening of a letterpress or litho print by oxidation/polymerisation can be assessed by using a suitable rub-tester (page 210).

Lithographic performance

Various attempts have been made to develop laboratory tests from which it would be possible to predict the likely performance of an ink on a lithographic press. The Pope and Grey tester makes a fairly direct approach to the problem by having a roller system in which the bottom distributor roller can be immersed in a lithographic fountain solution. The tendency for the ink to bleed into this solution, for the formation of water droplets in the ink and for piling on the rollers due to the uptake of fountain solution, can all be assessed, but while it seems possible to identify inks with a poor lithographic behaviour, small differences between inks with a generally better performance are not readily shown up. At the present time there appears to be no substitute for the litho machine as a means of testing the performance of a litho ink.

Printability

The complex mixture of properties of ink and paper which go to make up their printability can only finally be studied on some type of printing machine, and these printability tests are only of value if they are carried out under standard conditions. The nearest approach to this is achieved by using a simple form of printing machine or by using one of the laboratory printing devices like the IGT Printability Tester. Once standard printing conditions have been established with a standard ink and standard paper, then the performance of an 'unknown' ink can be examined. It only remains to standardise the method of assessing the prints produced.

Ink film thickness

In carrying out printing tests it is useful to be able to measure the thickness of a wet ink film on a slab or on a printing roller. A number of devices have been developed to make this measurement. These either operate on a roller principle (Cheville, Interchem) or on lens indentation (PIRA). The PIRA ink-film thickness gauge (fig. 14.13) consists essentially of a low-powered-microscope, one lens of which is pressed into the wet ink film. The lens picks up a spot of ink – a round spot from a flat surface, an elliptical one from a cylinder. The spot is projected on to a grid of lines 0·1mm apart and the picture is viewed through a magnifying eyepiece. Tables relate the dimensions of the ink spot to the thickness of the ink film.

TESTS CARRIED OUT ON PRINTS

After an ink has been applied to paper or some other substrate and given time to dry, tests can be carried out on the print which give some indication of how effectively the print will perform during its working life, for example, as a soap carton, as a poster or as a banknote. These tests may be carried out by pigment and resin suppliers, by printing ink manufacturers, by printers and by some of their customers. Where the tests are being made on samples of ink, then they must be preceded by the preparation of a standard print.

BS 4321:1969 Method 1 gives details of a method of preparing standard prints of letterpress or lithographic inks on certain specified coated and uncoated papers.

Testing methods on prints may be conveniently divided into the following groups:

Optical Properties: Colour; gloss.

Resistance Properties: Rub-resistance; light-resistance; resistance to water, solvents, alkali, acid, soap, detergent, oils and fats, waxes.

Colour and gloss

Since print is produced to be looked at, it is hardly surprising to find that the colour and gloss of a print is usually visually assessed against some agreed standard. Whether they are produced on a printing machine, by drawdown or by hand roller, it is obviously important that the ink film thicknesses of the two samples are approximately the same. There are occasions when a more permanent and precise record of colour or gloss is required, and suitable measuring instruments are listed on pages 259 and 260.

Rub-resistance

The ability of printed matter to stand up to abrasion is particularly important in its packaging applications, when the print may first be rubbed against metal parts on package making or package filling machines and then later rubbed against the surface of other print in the transportation of filled packages in outer containers. Tests in which one print is rubbed against another by hand are not satisfactory, because it is impossible to standardise the pressure and the speed of the rub. A number of instruments have been developed to measure rub-resistance and the British Standard BS 3110:1959

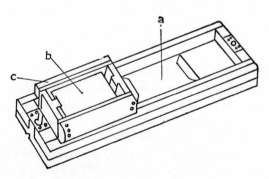

14.15 A simple rub tester (Proctor and Gamble).

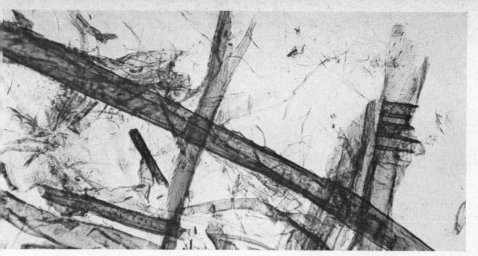

Plate 1. Photomicrograph of spruce fibres—mechanical pulp (magnification × 144).

Plate 2. Photomicrograph of spruce fibres—bleached sulphite pulp (× 144).

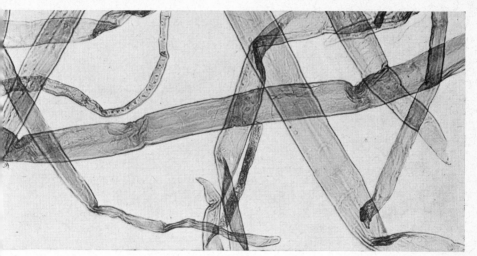

Plate 3. Photomicrograph of spruce fibres—bleached kraft pulp (× 144).

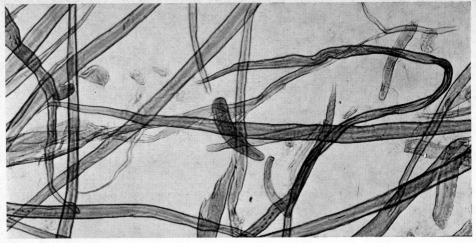

Plate 4. Photomicrograph of mixed species of bleached hardwood fibres—neutral sulphite pulp (×144).

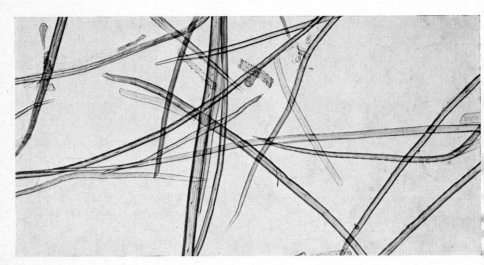

Plate 5. Photomicrograph of esparto fibres (×144).

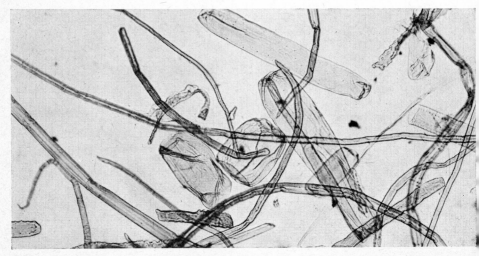

Plate 6. Photomicrograph of straw fibres (×144).

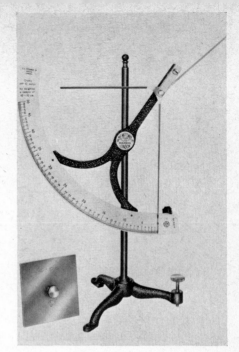

Plate 7. Quadrant balance. The template is used to prepare a standard sheet (10 cm × 10 cm) which is carried on one arm of the balance. The substance of the paper in grammes per square metre is indicated by the other arm moving across the quadrant scale.

Plate 8. Deadweight dial micrometer. A standard pressure is exerted on sheets placed in the jaws and the dial reading gives a measure of thickness.

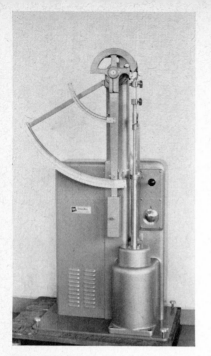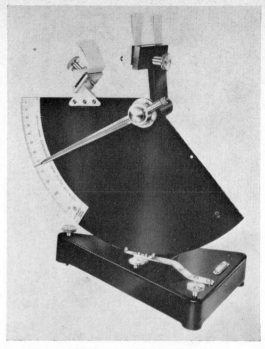

Plate 9 (left). Schopper tensile tester. A gradually increasing load is applied to the test strip held between the two clamps. The weighted pendulum moves across the quadrant scale and is held by a rachet at the point when the paper breaks, to indicate the tensile strength of the strip.

Plate 10 (right). Elmendorf tear tester. The tearing of a standard paper sample acts as a break on the pendulum swing of a heavy sector-shaped plate, preventing it from reaching the same height on the other side. The extent of which the plate falls short of that height is indicated by a pointer on a scale round its circumference, and this provides a measure of a paper's tearing resistance.

Plate 11. Mullen burst tester. A paper sample is firmly clamped by a ring over a rubber diaphragm. Pressure developed behind the diaphragm forces the paper out until it bursts. A gauge measures the pressure at the moment of burst.

Plate 12. PIRA Stack Thermometer. The temperature inside a ream or pallet of paper is measured by breaking the outer wrapping and inserting the long stem of the stack thermometer.

Plate 13. Printsurf roughness tester. The instrument provides information on the roughness of a paper by measuring the rate at which air under pressure escapes under a metal ring held on the sample, backed by an actual press packing or a standard equivalent. Measurements are made under printing pressures.

Plate 14. IGT Printability tester—distribution unit. The unit consists of a simple system of two steel rollers and one polyurethane roller to distribute a measured volume of ink and ink up a small metal disc. The instrument shown has a duplicate set of rollers, to allow the study of wet-on-wet printing.

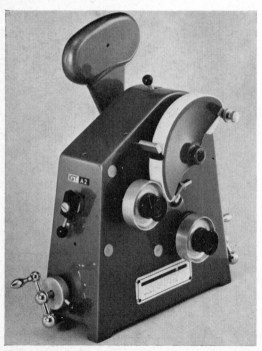 

Plate 15 (left). IGT Printability tester—printing unit. Ink is transferred by the small metal disc to a strip of paper clamped round the 150° impression sector, the counterpart to the impression cylinder of a machine.

Plate 16 (right). NPIRI—Chatillon tack finger. A small recessed circular plate on top of the instrument is inked up using a drawdown technique, the operator places his finger on the inked surface and then raises it as rapidly as possible. After repeating the film splitting process a specified number of times the maximum force is read directly on the dial gauge, to give a measure of the tack of the ink sample.

Plate 17. PIRA drying time tester. Ink is rolled out on to a circular glass plate which is mounted on a turntable driven by a motor. As the disc rotates a stylus draws a spiral track on the wet film of the ink. As long as the film is wet the track has smooth edges, but a point is reached when the ink is sufficiently dry for the film to be torn by the stylus, leaving a track with jagged edges.

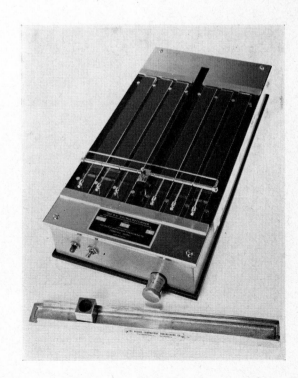

Plate 18. BK drying time recorder. A chromium-plated carrier draws six hemispherical needles along six test strips, so recording the drying characteristics of the coatings against the time scale, which runs along the length of the instrument The graduated knob on the front of the recorder allows a choice of three different speeds of travel.

Plate 19. PIRA rubproofness tester. Samples of the print are mounted on to two discs of different diameters, which are rotated in contact with one another.

Plate 20. PIRA print drying time recorder. A long strip of each print is mounted parallel to the axis of a slowly rotating drum and covered by another strip of unprinted paper. As the drum rotates at the rate of one revolution every nine minutes, a weighted stylus slowly moves along the drum so that every nine minutes the unprinted paper is pressed against another small portion of the drying print. As long as the print is wet, a set-off mark appears on the unprinted paper. These marks join up to form a solid band which only ends when the print is dry enough not to set-off, and the length of the band is a measure of the drying time of the ink on the paper.

'Methods for Measuring the Rub-Resistance of Print' describes three of these instruments (Proctor and Gamble, PIRA, Sutherland), each making different approaches to the property being measured. The first of these is a simple hand reciprocating test instrument developed by Proctor and Gamble in order to specify the rub-resistance required for their soap and detergent packs produced in the UK. It consists of a simple wooden frame (a), providing a channel in which a block (b) of standard dimensions and weight can be moved backwards and forwards by hand (fig. 14.15). Samples of the print under test are clamped on to the frame and on to the bottom of the sliding block. In the PIRA instrument (plate 19) samples of the print are mounted on to two discs of different diameters, which are rotated in contact with one another. In the Sutherland Rub Tester, one sample of the print attached to a rectangular metal block is drawn over the surface of a second sample in a constantly repeated arc. Details of the three instruments may be found in BS 3110 or by reference to the sources listed on page 258.

Light-resistance

Posters, showcards, and other display materials all need to have reasonable resistance to light. Although no print can be completely permanent, the ability of different inks to stand up to prolonged exposure varies considerably. In general this is due to the different lightfastness of the pigments in the inks, but it has been shown that the light resistance of a pigment does depend on the vehicle in which it is dispersed and on the % of pigment present.

Tests in which prints are exposed to daylight are time consuming and awkward to carry out, but they do represent the nearest approach to the conditions which the print will meet in practice. BS 4321 Test 2 sets out a standard method for daylight testing in which prints are mounted in exposure racks alongside samples of dyed wool cloth of known light fastness. The prints are protected from the weather by glass, not less than 50mm from the specimens. Accelerated tests for light fastness using intense light sources, *eg* mercury vapour or enclosed carbon arc lamps, are much more convenient than daylight exposures but they can give misleading results because the short wave ultra violet waves given out by these sources, although not present in daylight, fade inks fairly rapidly. High pressure xenon lamps have been found to give results most similar to those using daylight, and a standard method of determining the light resistance of prints using a xenon lamp is set out in BS 4321 Test 2A.

Resistance to water, solvents, and other materials

Depending on their function, printing ink films may come in contact with various substances during their working life, and it is important that they should be resistant to the materials which they are likely to meet. For example, a soap wrapper should be resistant to water, alkali and soap, a label for a vinegar bottle should be resistant to acid. Simple tests have been devised to measure the resistance of dry prints to a variety of materials. Fuller details of the tests which follow may be found in BS 4321:1969 'Methods of test for printing inks'.

Water

The print is pressed between two sheets of moist filter paper by two glass plates under a 1 kg weight for 24 hours. Changes in the appearance of the print and the bleeding of colour on to the filter paper are then assessed.

Solvents

Strips of the print are placed in the solvent (eg industrial methylated spirits) contained in a test tube at 20°C. The strips are removed after 5 minutes and dried at 40°C for 20 minutes. The print is then compared with an untreated specimen, and the solvent in the tube compared with fresh solvent in a similar tube.

Alkali

The print is pressed between two sheets of filter paper, saturated with a 5% solution of sodium hydroxide (caustic soda), by two glass plates under a 1 kg weight. After 10 minutes the print is removed, rinsed in water until neutral to an indicator (phenolphthalein), dried in an oven at 40°C for 45 minutes and assessed for change in comparison with an unprocessed print. The filter paper is also dried and examined for bleed from the print.

Acid

The acid resistance of a print may be assessed by substituting a 5% solution of hydrochloric acid for the 5% sodium hydroxide solution in the test outlined above. This test is not included in BS 4321.

Soap

The print is pressed between two filter papers, saturated with a 1% solution of the soap, by two glass plates under a 1 kg weight. After 3 hours the print is removed, rinsed, dried at a temperature of 40°C for 45 minutes and assessed for change in comparison with an unprocessed print. The filter paper is also dried and examined for bleed from the print.

Detergent

The method is similar to that outlined for soap except that a 3% solution of detergent is used in place of the 1% soap, and the time of the test is reduced from 3 hours to 1 hour. It is recommended that the name, date and source of the detergent should be quoted with the results.

Oils and fats

For fats which are solid at 20°C a specimen of the print is placed face down on a flattened sample of the fat in a Petri dish. After 24 hours the specimen is examined for colour change and the surface of the fat examined for bleed. Oils which are liquid at 20°C are applied on filter paper between glass plates as in the method described for soap.

Waxes

The print is immersed in molten wax at a temperature not more than 20°C above the melting point of the wax used. After 5 minutes the specimen is removed and the wax allowed to drip through the white margin of the paper into a basin. The print is examined for any colour changes and the wax on the white margin and in the basin examined for bleed.

Other tests on prints

BS 4321 also describes standard methods for measuring the resistance of prints to cheese and spices. Many other tests that are in everyday use have been devised to assess the resistance of a print to the particular product being packed, *eg* cat food, chutney. Inks that are going to be stoved dried should be tested for heat resistance.

Following a study of the subject of toxicity and food safety requirements the Society of British Printing Ink Manufacturers published a report on 'Printing Inks for Food Wrappers and Packages', which was revised and re-issued in 1969. The report defines four categories of ink use on packages varying in potential risk and it makes clear recommendations on the exclusion of all substances known to be toxic from inks for immediate wrappings, *ie* wrapping material with which the food is in direct contact. A list of materials that should be excluded from the formulations of inks designed for immediate food wrappers has been agreed between the Society of British Printing Ink Manufacturers, the British Food Manufacturers and the Food Research Association. Tests for the presence of lead and other toxic metals may either be carried out on a sample of the print or on separate samples of the ink and paper.

15. Light-sensitive materials

Certain materials undergo a physical or chemical change when they are exposed to light. To the printer, this may mean that the brilliant white dress on the poster, advertising a certain brand of washing powder is unfortunately transformed to a dull grey colour on the hoardings, but on the credit side it also means that he has a range of remarkable materials available for the processes of platemaking. These light-sensitive materials fall into two main classes, firstly the normal photographic materials based on the silver halides, and secondly a range of sensitive materials which are used in transferring a photographic image on to a printing surface in producing letterpress, litho or gravure plates.

PHOTOGRAPHIC MATERIALS

The printing industry is becoming more and more dependent on the processes of photography and with the rising demand for colour and the swing to filmsetting, one of the clearest things to be seen in the printing industry's crystal ball is that this trend will continue into the future.

The photographic emulsions, which are coated on to film, paper and glass, basically consist of microcrystals of certain silver salts dispersed in gelatin. These salts form the light-sensitive component of the photographic layer, and on exposure to light, they begin to release a finely divided form of metallic silver, which is black. The process begun in the camera is multiplied several million times in the process of development to produce a negative or positive, whose black areas consist of this finely divided silver.

Silver halides

Most silver salts are photosensitive but three have been found to be outstandingly suitable for photographic purposes: *silver chloride* which is used for slow (contact) printing papers, *silver bromide* which is used in negative emulsions and *silver iodide*, small quantities of which are generally included with the bromide in fast negative emulsions. Like the halogens, the silver halides form a family of compounds having similar properties and with any differences in these properties tending to be stepped down the group (Table 15.1).

Table 15.1 Three members of the silver halide family

Silver chloride	AgCl	white	All relatively insoluble but	
Silver bromide	AgBr	pale yellow	solubility increasing up the	Sensitivity to light increasing
Silver iodide	AgI	lemon yellow	table	down the table.

Silver halides can be prepared simply by mixing solutions of a soluble silver salt, normally silver nitrate and a soluble alkali halide. For example, silver bromide, the most important of the three halides photographically, is precipitated when a solution of potassium bromide is added to a solution of silver nitrate.

$$AgNO_3 \ + \ KBr \ \longrightarrow \ AgBr \ + \ KNO_3$$
soluble soluble insoluble soluble

When prepared in this way the silver bromide particles coagulate and settle out of solution in coarse lumps which would be useless in a photographic coating. In making a photographic emulsion, it is necessary to obtain the silver halide in the form of small particles evenly suspended in a solution. This is just one of the many functions of gelatin in photography.

Gelatin

Gelatin is a protein obtained by the hydrolysis of animal bones and tissue. Although continuous research efforts have been made to find a synthetic substitute, gelatin's position in photography has never been seriously challenged. It owes this position to a remarkable combination of properties, each one contributing to the performance of an emulsion. It has even been suggested that God must have had photography in mind when he invented the cow.

Gelatin's excellent adhesion to many surfaces is well known, because animal glue is an impure form of gelatin (page 233). Its films are extremely tough and so they are able to provide good mechanical protection for the silver halide grains. Gelatin swells in cold water but does not dissolve, hence it is possible for processing solutions like developers and fixers to reach the grains without removing the layer in which they are embedded. It dissolves in warm water at about 35°C to form a colloidal dispersion. If a silver halide is precipitated in the presence of this dispersion, the colloidal particles of gelatin are able to carry the separate grains of the halide in an extremely fine form and so act as a 'protective colloid'. When a gelatin solution is cooled it sets to a 'gel', providing that more than about 2% of gelatin is present in the solution. This easy transition from the 'hydrosol' to a 'hydrogel' provides a convenient means of setting a photographic emulsion when it is coated on to a substrate. Another useful property of gelatin is that it can be hardened by such chemicals as formalin or alum, to reduce its tendency to swell in water and increase the temperature at which it dissolves in water. A further advantage is that ordinary samples of gelatin contain minute quantities of certain chemicals which greatly improve the sensitivity of photographic emulsions. These impurities can be removed to produce inert gelatins and synthetic sensitisers may then be added in controlled quantities to produce emulsions with particular characteristics.

Photographic emulsions

We have seen that silver bromide is precipitated when silver nitrate is mixed with a solution of potassium bromide. If a solution of gelatin is added to the potassium bromide before pouring in the silver nitrate, a photographic emulsion is produced. This sounds very simple but in fact, emulsion technology is an extremely complex subject, which could fill another book. The size and distribution of the halide grains have an important influence on the sensitivity of an emulsion, and so the conditions for the precipitation reaction, *eg* solution temperatures and concentrations, speed of addition, must be carefully controlled. Furthermore, commercial emulsions contain small quantities of a number of other additives, including such materials as glycerine to make the dried layer more flexible, an alcohol to reduce froth formed during coating, surface active agents to make the film more easily wetted by subsequent processing solutions, hardening agents to decrease the swelling and solubility of the gelatin, gold salts and other sensitisers to improve the light sensitivity and dyes to adjust the colour sensitivity of the emulsion.

Photographic emulsions are coated on to film, paper or glass, the film normally being cellulose acetate, polystyrene or a polyester (pages 41, 45, 46). A high speed emulsion layer is about 10μm thick and contains of the order of 5×10^8 grains per square centimetre in a size range between 0·2 and 4·0μm. The proportion of silver halide to gelatin by weight is about 40/60 for an average negative emulsion. This proportion is much higher for a lith emulsion and reaches an upper limit of about 80/20 in a nuclear track emulsion.

Exposure

When a silver halide is exposed to light it decomposes into its constituent elements, metallic silver and a halogen (chlorine, bromine or iodine). If the exposure is extremely prolonged, the amount of silver formed is sufficient to give a visible darkening of the halide grains. This property forms the basis of 'print-out' silver processes. However, the shorter exposures, which are more normal in photography, only release a small number of silver atoms and there is no visible darkening in the areas receiving light. The small clusters of silver atoms on the halide grains constitute a 'latent image' which can be amplified many millions of times in the subsequent process of development. It has been calculated that for every atom of silver formed in the camera, 1 000 000 000 are formed in the developing bath. Exposure is to development rather as a detonator is to a bomb. The second stage gives much the greater effect, but it is not possible without the first.

When an exposed emulsion is developed, the black image starts its growth from one or two isolated specks or 'development centres'. These centres, which in aggregate constitute the latent image, consist of small groups of silver atoms. The theories of latent image formation produced by a succession of photographic scientists set out to explain the steps following the absorption of light quanta by a grain which finally lead to the formation of small groups of silver atoms on the developable grain. Although these theories are still being refined, they owe much to the basic ideas put forward by the British scientists Gurney and Mott in 1938. The silver and halide ions in a photographic grain are arranged in the form of a cubic lattice, each silver ion having six halide neighbours and each halide ion having six silver neighbours (fig. 1.5). According to Gurney and Mott the photochemical reaction leading to the formation of a latent image takes place in four stages. First, a quantum of light falling on a negatively charged halide ion causes an electron to be released into the lattice. This electron moves very rapidly until it comes in contact with one of the local areas of dislocation in the crystal lattice. In the second stage the electron is captured by a sensitivity centre, which then becomes negatively charged. In the third stage a slow moving free silver ion known as an 'interstitial ion' is drawn through the lattice towards the negatively charged centre where it is neutralised to form one atom of silver. In the fourth and final stage the halogen atom, formed initially on the absorption of light, escapes from the lattice and combines with the surrounding medium. If the process is repeated so that the number of silver atoms in the group reaches a certain number then the grain becomes developable. It is important to realise that the actual exposure necessary to produce a developable grain does vary with the nature of the developer solution and with the temperature and time of development.

Development

The processes of exposure and development convert silver ions (Ag^+) into silver atoms (Ag).

Ag^+ + one electron → Ag

silver ions in the   metallic silver
silver halide crystal   in a finely divided
lattice   form
(colourless)   (black)

Since this change involves the acquisition of an electron and is therefore a chemical reduction, the principal ingredient in a developing solution is a substance capable of donating electrons, *ie* a reducing agent. Although there are many thousands of reducing agents, a comparatively small number of them will selectively reduce only the light-struck grains. Reducing agents which are too active will reduce all the silver grains, exposed and unexposed alike, whilst many weak reducing agents will not even affect the exposed grains. The successful developing agents in common use are generally derivatives of benzene in which two or more hydrogens have been replaced by hydroxy ($-OH$), amino ($-NH_2$) or substituted amino groups (*eg* $-NH.CH_3$). The formulae of the three commonly used reducing agents in developer are given below.

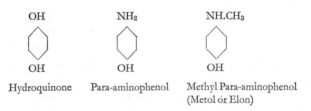

OH	NH$_2$	NH.CH$_3$
OH	OH	OH
Hydroquinone	Para-aminophenol	Methyl Para-aminophenol (Metol or Elon)

Since these reducing agents are only active in alkaline solutions, developer baths normally contain sodium hydroxide, sodium carbonate or some other alkali acting as an *accelerator*. A secondary effect of this alkali is to increase the swelling of the gelatin which is at a minimum at the iso-electric point (pH 4·8). One of the products of the development reaction is potassium bromide. Since in any chemical reaction, the addition of a product has the effect of slowing up a reaction, potassium bromide may be added as a *restrainer*, controlling the rate of development. There is a tendency for oxygen in the air to oxidise the developing agent, and normally sodium sulphite is included as a *preservative* preventing this atmospheric oxidation and so keeping the solution clear.

Fixing the image

Those areas which have not been reduced to metallic silver, still carry a coating of unexposed silver halide. This coating remains light sensitive and if left would slowly darken and fog the whole photograph. The process of removing these unexposed grains is called *fixing*. Since the silver halides are insoluble in water, use has to be made of the fact that they form soluble coordination compounds with solutions of salts containing thiosulphate or cyanide ions. Sodium thiosulphate (or Hypo, $Na_2S_2O_3$) is invariably used, the probable reactions being as follows.

$$AgBr \ + \ Na_2S_2O_3 \ = \ NaAgS_2O_3 \ + \ NaBr$$
insoluble   soluble   slightly soluble   soluble

$$AgBr \ + \ 2Na_2S_2O_3 \ = \ Na_3Ag(S_2O_3)_2 \ + \ NaBr$$
insoluble   soluble   soluble   soluble

There is a tendency for developing solution to be brought over into the fixing bath, thus allowing development to continue unless it is checked. One answer to this is to rinse the photograph in a bath of water or weak acid between developing and fixing. However, a more usual way is to use an acid fixing bath. Acidic conditions effectively stop development, since the reducing agent is normally active in alkaline solution. Unfortunately, hypo is not stable in acid solutions and sulphur is formed, giving a cloudy bath.

$$Na_2S_2O_3 \ + \ 2HCl \ \longrightarrow \ H_2SO_3 \ + \ S \ + \ 2NaCl$$

This reaction can be reversed by the presence of a high concentration of sulphite ions. The sulphite ions together with acidic conditions are usually provided either by sodium metabisulphite or by sodium sulphite and acetic acid. Another common ingredient in a fixing bath is a hardener, *eg* potash alum, which has the effect of reducing the swelling of gelatin in subsequent processing.

Aftertreatments

After developing and fixing it may be necessary to modify the density of a negative, and various methods are available to reduce or intensify the silver image.

Photographic reduction is achieved by converting some of the metallic silver to a silver salt, and dissolving this away in a suitable solvent. For example, the commonly used Farmers reducer contains potassium ferricyanide which converts silver to silver ferrocyanide, and 'hypo' which dissolves this away.

$$4Ag \; + \; 4K_3Fe(CN)_6 \; = \; Ag_4Fe(CN)_6 \; + \; 3K_4Fe(CN)_6$$

4Ag	+	4K₃Fe(CN)₆	=	Ag₄Fe(CN)₆	+	3K₄Fe(CN)₆
Silver		Potassium ferricyanide		Silver ferrocyanide		Potassium ferrocyanide
				↓		
				dissolves in 'hypo'		soluble

Farmers reducer lowers all densities by an equal amount and so proportionally has the greater effect on the lower densities. Subtractive reducers of this type are most suitable for those negatives which have been correctly developed to the right contrast but have a high density due to gross over-exposure. Other types of reducers known as proportional reducers react roughly in proportion to the amount of silver locally present in the image, so that all densities are reduced by the same proportion. This lowers the contrast of a negative as well as its general density and it may be used when a negative has been correctly exposed but over-developed. The reduction in the blackness or optical density of a silver image should not be confused with chemical reduction. In fact, photographic reduction is always achieved by chemical oxidation.

The density of a silver image may be intensified either by physical or chemical means. In *physical intensification*, metallic silver or mercury is deposited on the silver image by immersing the negative in a solution containing a chemical reducing agent, *eg* a developer, and a silver or mercury salt. In *chemical intensification* the silver image is first converted to a silver halide by an oxidising agent which is itself reduced to an insoluble compound, and then both substances are changed into a black metallic form. For example, mercuric chloride reacts with silver to form silver chloride and mercurous chloride, both of which can be converted to the metallic state by ammonia or a developer.

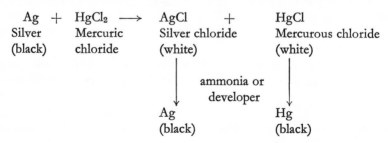

Ag	+	HgCl₂	⟶	AgCl	+	HgCl
Silver (black)		Mercuric chloride		Silver chloride (white)		Mercurous chloride (white)
				↓ ammonia or developer		↓
				Ag (black)		Hg (black)

Lith emulsions

These materials, now widely used in graphic reproduction, are noted for their ability to give high contrast and image sharpness. They differ from most other emulsions in that they are based on silver chloride and they require a special developer containing hydroquinone and formaldehyde.

OTHER LIGHT-SENSITIVE COATINGS

The production of a printing plate for each of the major printing processes essentially involves the conversion of a photographic image on film or glass into a relief, intaglio or lithographic image on metal or plastic. This is usually achieved by making use of the change in solubility of a light-sensitive coating on exposure to light. If this light-sensitive coating is exposed through a negative or positive, the areas receiving light usually become harder and less soluble. These areas may then be used as resist areas in etching the metal surface of the block, plate or cylinder, or they may actually form the basis of a litho image. Although there are many differences in practical detail for each process, this is basically the method by which the image on a printing plate for letterpress, litho and gravure is produced from a photographic image.

The light-sensitive coatings used at the printing down stage of platemaking fall into a number of chemical classes, the most important being

(a) *Dichromated colloids*, each consisting of a dichromate salt combined with a natural or synthetic polymer, *eg* potassium dichromate and gelatin, ammonium dichromate and polyvinyl alcohol.

(b) *Diazo compounds*, synthetic organic chemicals used particularly in producing presensitised litho plates.

(c) *Photopolymers* which may either be materials in which the polymerisation of small molecules (monomers) is triggered by the action of light, *eg* 'Dycril', 'Nyloprint', or unsaturated long chain polymers which on exposure to light crosslink to form larger molecules, with a lower solubility, *eg* 'Kodak Photo Resist'.

(a) Dichromated colloids

Until comparatively recently, mixtures of dichromate salts and certain natural polymers were the only light-sensitive coatings used in platemaking Over the years a great variety of natural polymers have been tried successfully, but the most commonly used have been gum arabic, gelatin, fish glue and egg albumen. They are known as colloids because they form colloidal dispersions in water.

Gum arabic is obtained from certain varieties of the acacia tree grown in the

Middle East. It is collected from the trees, separated from the bark, dried, graded and supplied in the form of yellowish brown lumps. These slowly dissolve in water to form a gummy colloidal dispersion. Gum arabic sensitised with ammonium dichromate is used in the production of 'deep etch' litho plates.

The protein *Gelatin* has been considered in some detail earlier in this chapter. When sensitised with potassium dichromate it is used in producing plates and cylinders for gravure. In other platemaking processes, the dichromated colloidal dispersion is coated direct on to the metal, dried, exposed and developed, but in gravure the situation is rather different. Carbon tissue, a thin paper coated with pigmented gelatin, is sensitised by immersion in a dilute solution of potassium dichromate. This is then laid face downwards on to plate glass, dried, exposed through screen and positive, and developed before being transferred to the copper plate.

Fish glue or gelatose is a degraded form of gelatin obtained from fish skin and bones. Gelatose differs from gelatin in that it will never form a gel. It is supplied in the form of a thick colloidal solution which is mixed with ammonium dichromate and more water to form the coating solution. Dichromated fish glue was used extensively to produce halftone blocks for letterpress, but its place has largely been taken by *polyvinyl alcohol*, a water-soluble synthetic polymer. The exposed areas of these coatings need to be baked before they form an effective acid resist for the etching of blocks, but the 'burning-in' temperature is lower for polyvinyl alcohol than for fish glue. Polyvinyl alcohol also has the advantage that its chemical composition and properties can be more effectively standardised from batch to batch than is possible with natural materials like gelatose.

Egg albumen is another protein which has found considerable use in platemaking. Its colloidal dispersions in water may be sensitised with ammonium dichromate, to provide light-sensitive coatings in the production of letterpress line blocks and surface litho plates. Solutions of all these natural colloids are subject to bacterial attack and mould growth, but preservatives may be included to improve their stability.

The sensitising salt for these natural or synthetic polymers is either ammonium dichromate or potassium dichromate. Each have orange-red crystals which dissolve in cold water, but ammonium dichromate is considerably more soluble than the potassium salt. Potassium dichromate is preferred for sensitising gelatin for gravure platemaking.

The nature of the chemical change taking place when a layer of dichromated colloid is exposed to light has been studied by a number of research workers. It has been shown that the polymer is chemically oxidised and the dichromate reduced, the chromium in the dichromate molecule being converted to a trivalent form (Cr^{+++}), which is able to act as a bridging link between the colloid molecules. By this cross linkage, very large polymer networks form, which as one would expect are harder, tougher and less soluble than the original colloid molecules.

A number of factors affect the rate of hardening of a dichromated colloid. These include the amount of incident light, the spectral characteristics of that light, the temperature and relative humidity, the ratio of dichromate to colloid, the pH and the thickness of the coating. In industrial practice it is important that each of these variables is kept under reasonable control.

Natural polymers like gelatin and gum arabic, sensitised by a dichromate salt, have been successfully used since the earliest days of photomechanical reproduction, and they remain important today. However, these materials do have a number of serious drawbacks: they are relatively insensitive by normal photographic standards, requiring long exposures to ultra-violet or visible light of short wavelength; the hardened coating may require an additional 'burning-in' process before it forms an effective etching resist; the hardening reaction takes place to some extent in the dark so that dichromated colloid coating mixtures cannot be stored for long periods before use; dichromate salts are inclined to be dermatitic and so introduce a health hazard; finally, as with all natural materials, it is difficult to control quality. These drawbacks form the background to the efforts made to replace dichromated colloids, and to the applications of diazo compounds and photopolymers in platemaking.

(b) Diazo compounds

Diazo compounds form a family of man-made chemicals all of which contain two linked nitrogen atoms ($-N_2-$). They are prepared from compounds called amines, containing the $-NH_2$ group. When these amines are treated with nitrous acid, HNO_2, in the presence of a second acid, usually HCl, a 'diazotisation' reaction takes place and a diazo compound containing the $-N_2-$ group is produced.

If we represent our amine as $R.NH_2$ then the diazotisation reaction may be summarised in the general equation

$$R.NH_2 + HNO_2 + HCl \longrightarrow R.N_2Cl + 2H_2O$$

One of the best known amines is the substance aniline, $C_6H_5NH_2$, a close relative of the solvent benzene, C_6H_6, both compounds having their carbon atoms linked in a hexagonal ring arrangement. When aniline is diazotised with nitrous acid and hydrochloric acid the $+NH_2$ group is replaced by N_2Cl in the diazo compound which forms.

Diazo compounds are very reactive substances and so tend to be unstable. Their instability to heat means that the diazotisation reaction normally has to be carried out near 0°C in a vessel cooled by ice. They are also unstable to the action of light, and it is this property that is made use of in many of their industrial applications.

The action of light causes the diazo group $—N_2—$ to break off and be replaced by the hydroxyl group $—OH$. For example, the diazo compound prepared from aniline (see above) is converted into phenol.

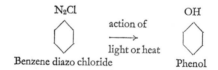

In every case, the resulting compound has very different properties from the original diazo compound. As far as platemaking applications are concerned, the most important change is that the compound formed is normally much less soluble than the diazo compound.

Diazo compounds were first used in litho platemaking just before the 1939 war, as replacements for dichromate. On exposure to light the $—N_2—$ group of the diazo compound was replaced by an $—OH$ group, and the resulting compound had the effect of hardening and reducing the solubility of the albumen coating in which it was incorporated.

Later it was realised that some of these diazo compounds could be effective on their own and not simply as hardeners for albumen and other colloids. In other words, when the diazo compound decomposes under the action of light, the product formed is sufficiently resinous to act as the image area on a litho plate. Diazo materials of this type are still extensively used for coating negative-working pre-sensitised litho plates.

These compounds have quite large molecules but the diazo groups ($—N_2—$) attached to them make them soluble in water. As we have seen already the action of light removes these groups and so destroys the solubility of the compound. In fact, after exposure the compound is able to perform effectively as a litho image. The possible structure for one such diazo compound and the change taking place on exposure to light is shown in fig. 15.1.

CH_2O
formaldehyde

+

diazodiphenylamine

diazo condensate
(soluble)

diazo group
removed (insoluble)

15.1 The reaction taking place when a diazo condensate is exposed to light.

Another important property of diazo compounds is their ability to link up with certain other compounds (couplers) to form brightly coloured dyes or pigments. These azo dyes and pigments form the largest group of synthetic colouring matters used in the colouring of textiles, plastics and printing inks (page 162).

(c) Photopolymers

Polymerisation is the reaction in which large numbers of small molecules (monomers) link together in long chains to form giant molecules (polymers). Polymerisation and the polymers used in the printing industry are described in Chapter 4.

In a number of chemical processes it is possible for light to act as a trigger or detonator to start the reaction. For example, photography is concerned with the conversion of a silver salt into finely divided black metallic silver on the areas of the film receiving light. The short exposure to light in the camera starts the reaction but the small number of silver atoms formed is increased about a thousand million times in the subsequent process of development. In a similar way we find that for some monomers the process of polymerisation can be triggered off by the action of light. Once the process has started, a chain reaction takes place and the monomer molecules continue to link up until polymerisation is complete. Polymerisation which is initiated by the action of light is called photopolymerisation. Fig. 15.2 shows the structure of a simple acrylamide monomer, which on exposure to ultra-violet or visible light of short wavelength polymerises to form a polyacrylamide plastic.

Clearly the idea of photopolymerisation is very attractive in platemaking. Dichromated colloid coatings are relatively insensitive to light by photo-

Acrylamide monomer

repeating unit of polyacrylamide

15.2 Photopolymerisation of a simple monomer.

graphic standards. Photopolymerisation introduces the possibility of short exposures at the printing down stage of platemaking, because once the small amount of light energy has been supplied to start polymerisation, the process will continue under its own impetus. Another attraction of the process is that potentially there should be a great difference in solubility between the un-exposed and the exposed polymerised areas, so allowing an easy process of development.

In practice, however, there are a number of problems in using photo-polymers in platemaking. Since by the nature of things monomers are small molecules, they tend to be either gases or liquids. Neither form is convenient to handle in making a plastic printing plate nor for coating a light-sensitive layer prior to printing down. One method of overcoming this problem is to incorporate the monomer liquid into a second polymer material. On exposure to light the monomer both links with itself and also with the surrounding polymer, so changing the solubility of the mixture. Also included with the monomer is an initiator, a type of catalyst which improves the efficiency of the process in which light starts the polymerisation reaction.

These principles are said to be incorporated in the Dycril process in which it appears that an acrylamide monomer is incorporated in an alkali-soluble form of cellulose acetate. After exposure the polymerised areas are no longer soluble in the alkali (dilute sodium hydroxide) used to develop the plate. In the Nyloprint photopolymer platemaking system the polymer surrounding the light-sensitive monomer is an alcohol-soluble form of nylon. In this case the polymerised areas lose their solubility in the alcohol/water solution used in development.

The principles of photopolymerisation are also exploited in another group of materials used in photo-engraving and litho platemaking. These are long chain polymers, whose molecules are able to cross-link under the action of light to form a three-dimensional molecular network. The original polymer is insoluble in water but like most thermoplastic long chain polymers it is soluble in some of the common organic solvents. After exposure to light, and the formation of bridging links between the chains, the polymer ceases to be soluble except in the very powerful solvent mixtures of the types used in paint stripping.

$$-CH_2-CH\text{---}CH_2\text{---}CH-CH_2-$$

```
    — CH₂ — CH——— CH₂ ——— CH — CH₂ —
            |                 |
            O                 O
            |                 |
            C = O             C = O
            |                 |
            CH                CH
            ||                ||
            CH                CH
            |                 |
          (benzene          (benzene
           ring)             ring)
```

15.3 Part of a molecule of polyvinyl cinnamate.

One of these polymers, polyvinyl cinnamate, has a structure similar to that of polyvinyl chloride (page 43), but in place of the chlorine atoms there are side chains including unsaturated regions (fig. 15.3). On exposure to light the double bonds between the neighbouring carbon atoms break and bridging links are able to form between the long chain polymers (fig. 15.4). Polymers like polyvinyl cinnamate have low sensitivity to light when used on their own, but the addition of various organic substances has made it possible to develop coatings which are several times faster than the normal dichromated colloid.

Materials of this general type are said to form the basis for the range of Kodak photosensitive resists. Stencils produced using these materials are highly resistant to all commonly used solutions and will withstand electro-plating and electro-etching processes. Similar light-sensitive resins are also applied in the so-called 'polymer-sensitised' negative working litho plates. On exposure, cross-linkage takes place on the image areas of the plate to form a hard but greasy surface. Solvent development removes the original long chain polymer from the non-image areas.

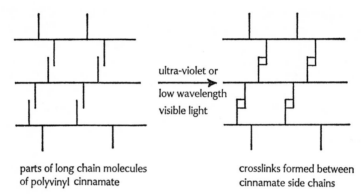

parts of long chain molecules
of polyvinyl cinnamate

crosslinks formed between
cinnamate side chains

15.4 Crosslinking of polyvinyl cinnamate.

16. Adhesives

It would be a slight exaggeration to say that without adhesives, modern civilisation would fall apart. Nevertheless they do have a vital, if often unseen, role to play in the construction of aircraft, cars, boats, buildings, furniture, and a thousand other things. Certainly, many of the products of the printing and packaging industries, including books, cartons, boxes, calendars and showcards, depend on adhesives. There are several different types of adhesives, and choosing the right one for a particular job is not always an easy task. It may involve a great deal of 'trial and error' although this can often be reduced if one has a basic knowledge of why things stick together.

THE NATURE OF ADHESION

Before considering the basic theories of adhesion we should first define some of the terms we will be using in this chapter. An *adhesive* can be considered as any material that causes one body to stick to another. The force holding two surfaces together is called the *force of adhesion*, and the materials being joined are known as the *adherends*. In considering the strength of a joint between two materials we are concerned both with the adhesion of the adhesive to the surface and also with the *cohesion* of the adhesive to itself. In other words, an adhesive must itself have good cohesive strength, as well as good bond strength to the adherends.

As yet, there is no general theory of adhesion which successfully explains all the facts. However, we can identify two quite different types of adhesion – one being due to forces of attraction between molecules of adhesive and adherend, and the other involving a mechanical linkage by an adhesive which penetrates into holes in the two adherend surfaces. This second type of 'mechanical adhesion' or interlocking can only apply when the two surfaces allow the adhesive to penetrate. It is rather like the bond obtained when a piece of plasticine is pressed between two pieces of wire gauze. Although 'mechanical adhesion' may contribute to the strength of a joint when porous materials, like most papers and boards, are being joined, the strength of the union between adhesive and adherend depends fundamentally upon the

229

forces of attraction between two sets of molecules. A proper understanding of why things stick together must be based on a knowledge of these molecular forces.

In the first chapter of this book we considered some of the ways in which molecules are built up by the formation of chemical bonds between atoms. We saw that the covalent bonds providing most of the linkages between atoms in organic substances are formed by the sharing of electrons between atoms, and that when there is an unequal sharing of these electrons, the molecules have a dipole moment. The positive and negative poles of these molecules are physically attracted to opposite poles on neighbouring molecules (fig. 1.8) and this molecular association affects such things as the melting point, boiling point and solvent power of these compounds (pages 8 and 172).

Physical forces of molecular attraction of this type are responsible for the vast majority of bonds across adhesive-adherend interfaces. They are called *van der Waals forces* and they fall into three classes.

1 *Keesom dipole-dipole forces* due to the attraction of two molecules, each having a permanent dipole moment.

2 *Debye dipole-molecule forces* where one molecule with a permanent dipole moment induces a temporary dipole in a neighbouring molecule which is normally non-polar.

3 *London molecule-molecule forces* which result from the polarisation of one molecule by another due to the oscillation of electron clouds.

These London forces become stronger with increasing molecular weight and this explains why polymers tend to have very good cohesive strength.

Of these three classes of inter-molecular forces, the Keesom and London forces are by far the most important. In order for Keesom forces to operate, an adhesive must be a polar compound, and in practice we find that most adhesives are organic compounds containing polar groups, eg gelatin, starch, polyvinyl acetate. Following this argument through, one might expect polar substances like acetic acid or ethyl alcohol to be good adhesives. The reason they are not is that molecules of this small size lack the cohesive strength which would be given by strong London forces. Just as the polarity of a solvent can be matched to that of the substance being dissolved (page 172), so the polarity of an adhesive can be matched to the polarity of the adherend. Table 16.1 lists some polar and non-polar materials which are either used in adhesives or are themselves adherends. Where an adherend lacks any polarity as with polythene film, adhesion can be made possible by treating its surface in some way to introduce polarity. To sum up we can say that most of the adhesives used in print finishing and packaging are polymeric materials containing polar groups. The strength of the adhesive bond is largely due to

Keesom forces operating between molecules of adherend and adhesive and the cohesive strength of the adhesive layer is due to London forces acting between the polymer molecules.

The first requirement for adhesion is that the adhesive should thoroughly wet the surface of the adherend. A high degree of wetting is indicated by a low contact angle (page 15), so the surface tension of the adhesive should be as low as possible. Just as we are careful to rub down and clean woodwork before painting, so it may be necessary to consider some form of surface preparation before the application of an adhesive. Roughening an adherend surface will assist 'mechanical adhesion' and provide a greater potential area of contact for adhesion to take place. Dust or grease on the surface will prevent adhesion, and these should be removed. The layer of oxide on a metal surface is not always as well anchored as that on aluminium, and it may be necessary to remove this oxide before adhesion can take place. Polyethylene, like a number of other synthetic polymers, has no dipole moment and is virtually impossible to polarise in its untreated state. Thus, it has no affinity for normal polar adhesives, and poses a difficult problem for the adhesive technologist. The only solution is to pretreat the polythene film by oxidising its surface with electrical discharge, gas flame or chemical oxidising agents in order to introduce polarity and reduce its contact angle with the adhesive.

If an adhesive is going to cover the full area of the joint and penetrate all the 'valleys' in the adherend's surfaces, it must have the ability to flow. The viscosity of the adhesive provides a measure of this ability and is one of the properties requiring close control in many industrial applications of adhesives. We have seen that in order to have good cohesive strength an adhesive should have a high molecular weight and normally it is a polymer. This requirement has to be combined with the need for the adhesive to be applied as a liquid at a fairly low viscosity. As with other surface coatings, a polymer layer can be applied as a low viscosity liquid in a number of ways. The solid polymer can be heated and applied to the adherend as a free flowing *hot melt*

Table 16.1 Examples of polar and non-polar materials

Polar	*Non-Polar*
Paper	Waxed paper
Metal oxides	Aluminium
Casein	Benzene
Starch	Polyethylene
Water	Mineral oil
Polyvinyl acetate	Polypropylene
Ethyl acetate	Carbon tetrachloride

adhesive. Alternatively the polymer may be applied as a solution in cold or hot water, as a solution in an organic solvent or as an oil-in-water emulsion. The control of viscosity is important for each of these types of adhesive. If the adherend is absorbent, too low a viscosity may lead to excessive penetration of the adhesive and insufficient will be left to form an adequate bond with the other adherend surface. On the other hand, if the viscosity is too high, there may not be enough penetration to produce an effective key and the spreading of the adhesive may be poor. Methods of measuring viscosity are discussed on page 201. The viscosity of a liquid varies greatly with its temperature, so any check on the viscosity of an adhesive should be made at the temperature at which it is going to be used on the machine. It should also be remembered that the viscosity of a solution adhesive varies with its solids content, and the viscosity of an adhesive on a machine will rise if water or some other solvent is allowed to evaporate. Unless corrective action is taken an adhesive which started with the correct viscosity may begin to cause trouble through poor distribution and a failure to penetrate the adherend.

TYPES OF ADHESIVE

Adhesives are produced from a wide variety of sources. Many of them are obtained from naturally occurring materials which may be animal, vegetable or mineral in origin. To these have been added a wide range of synthetic adhesives, whose development has been a part of the general growth in the industrial applications of synthetic polymers.

The more important classes of adhesives are shown below.

Animal glues
Fish glues
Casein adhesives
Starch-based adhesives including starch pastes and dextrine cold glues.
Natural resin adhesives including gum arabic, gum tragacanth and shellac.
Cellulose adhesives including those based on methyl, ethyl or carboxymethyl cellulose, nitrocellulose and cellulose acetate.
Rubber-based adhesives including natural and synthetic rubber latex and solution adhesives.
Inorganic adhesives including sodium silicate.
Synthetic resin adhesives including
thermoplastics (*eg* polyvinyl acetate, polystyrene, polyamides, etc) and thermosets (*eg* phenolics, epoxies)

NOTE *Hot melt adhesives* may be based on synthetic thermoplastic resins or they can contain paraffin and other waxes.

232

Adhesives from each of these classes find applications in the printing and packaging industries, although some of them are only used to a limited extent. Paper and board are the materials most commonly involved, and since their polar nature and generally porous structure are considerable aids to adhesion, the fairly simple water-based adhesives meet most needs. These may be the traditional animal glues and starch-based adhesives or the increasingly popular synthetic resin emulsion adhesives, like polyvinyl acetate.

Although these simple adhesives are adequate in the great majority of cases, difficult problems of adhesion can arise in packaging and printing, as in any other industry. For example, the area of the paper being glued may be covered with a hard film of printing ink or varnish. Alternatively the adherends may be completely non-absorbent plastic films. In cases like this, it may be necessary to consider some pre-treatment of the adherend surface and a more sophisticated synthetic resin adhesive may be required. When new problems of adhesion arise, a great deal of time and money can be saved by seeking specialist advice from one of the adhesive manufacturers.

This is not to suggest that contact with the adhesive manufacturers should be reserved for those occasions when difficult problems of adhesion have arisen. The choice of a particular type and grade of adhesive for any new job is not simply a matter of finding a substance which will form a good bond between the two adherend surfaces. Other major factors influencing choice are the cost of the adhesive, its working properties during the glueing process and the durability of the bond it produces. Among the working properties which may be of interest to the user are the adhesive's shelf life, its pot life after exposure to air, its viscosity, its setting time on a particular material and the rate at which the bond strength develops. It may be important for an adhesive for a particular job to be free from odour and from any toxic ingredients. Where the adhesive is going to come in contact with coloured paper, cloth or other similar material, the pH of the adhesive, should be such that the dyes or pigments present will not be discoloured. Clearly, finding the right adhesive for a given set of circumstances is not always a simple matter and there is everything to be gained by giving the adhesive supplier the fullest possible information on what will be demanded of an adhesive both during its application and afterwards.

Animal glues

Animal glues are known to have been in use over 3 000 years ago. They are obtained by treating animal skin, muscular tissues and bones with hot water or with milk of lime. In this process, the protein collagen passes into solution in a hydrolysed form which we call gelatin. Gelatin has been shown to consist of large molecules having a long polypeptide chain structure, built up from at least 18 amino acids (fig. 16.1). Its molecular weight ranges from 20 000 to 200 000 or more.

Amino acids   *eg* $H_2N—CH_2—COOH$ Glycine

$H_2N — R_1 — COOH$ $H_2N — R_2 — COOH$ $H_2N — R_3 — COOH$

$—H_2O$ $—H_2O$

linked
in a
Polypeptide chain

$H_2N — R_1 — CO — HN — R_2 — CO — HN — R_3 — CO — HN —$ etc

16.1 Chemical structure of gelatin.

Basically animal glue can be considered as an impure form of gelatin, although other substances are added in compounding the various commercial grades. Flexible glues always contain plasticisers like sorbitol or glycerol, which take up moisture from the atmosphere and so prevent the glue layer from drying out and becoming brittle. Animal glues are soluble in warm water and they are usually applied as solutions at about 60°C. They set rapidly by gelation on cooling. Heating glue solutions above 60°C should be avoided since it has the effect of degrading the gelatin. Animal glues have been widely used for many years in the bonding of wood, paper, leather and cloth, particularly in the furniture and other woodworking industries and in book-binding.

Fish glue

Fish glue, obtained from the skins of fish, is similar in properties and applications to animal glue. Dichromated fish glue was formerly important as a light sensitive coating in photo-engraving.

Casein

Casein is another protein adhesive used in bonding wood and to a lesser extent paper, leather and cloth. It is prepared by acidifying skimmed milk to a pH of 4·5 when the casein separates as a curd and is washed and dried. Casein glues are often supplied in powder form and require mixing with water before use, usually as cold press adhesives.

Starch-based adhesives

Starch-based adhesives are extensively used in glueing paper and board. Starch is a polymeric material found in the seeds, stems, leaves, roots and tubers of plants. Different varieties of starch are extracted from a number of plant sources including wheat, maize and potatoes. Like cellulose, starch is a carbohydrate and has a similar chemical structure of six-membered glucose rings linked together through oxygen atoms (fig. 16.3).

16.2 Part of a starch molecule.

The high proportion of hydroxyl groups give the polymer a strong affinity for water, in which it swells and forms a gelatinous solution. Pastes of untreated starches tend to be slow drying because of their low solids content, but by converting the starch into a degraded form (dextrin) faster drying adhesives can be produced with a lower viscosity coupled with a higher solids content.

Starch-based adhesives are supplied as cold water pastes or as powders to be mixed with water before use. They are light in colour and relatively inexpensive. Although their strength is low compared with other types of adhesives, it is perfectly adequate for many uses involving paper and board, eg bag making, carton and box manufacture, gummed envelopes, gummed tape. Starch and starch products are also used as sizing agents in the paper making process.

Natural resin adhesives

Most of the natural resin adhesives are solutions of resinous materials exuded from the bark of certain trees, eg gum arabic, gum tragacanth. These gums are of limited importance in the adhesive field, but gum arabic is of wider interest in the printing industry because of its application in lithography both as a desensitising material on the non-image areas of a plate and also in combination with ammonium dichromate as a light sensitive coating in deep etch litho platemaking. Although most natural resin adhesives are of vegetable origin, shellac is an important exception since it is produced by a parasitic insect living on certain trees in India. For adhesive purposes it is either used as a hot melt or in alcoholic solutions.

All the adhesives discussed so far are of natural origin and suffer from the disadvantage that they can become breeding grounds for bacteria, especially in warm and humid conditions. They also allow the tiny spores of moulds and fungi, small enough to get through most air filters, to develop into the large colonies which can sometimes be seen as patches of mould on bindings. Starch based adhesives provide a natural food for a range of insects, as well as mice, which will eat their way through book covers to reach the starch or dextrin in the adhesive layer. The addition of small amounts of various chemical preservatives in animal or vegetable adhesives greatly improves their resistance to attack by bacteria, moulds and insects.

Cellulose adhesives

Cellulose adhesives consist basically of solutions of chemically modified forms of cellulose. Some of the more important of these cellulose derivatives are described on page 40. They include materials like nitrocellulose and cellulose acetate, which are soluble in organic solvents and whose solutions are used as clear waterproof adhesives. The group also includes water soluble derivatives, like methyl cellulose and sodium carboxymethyl cellulose (CMC) which are used as non-staining wallpaper pastes, as surface sizing and coating agents in paper and board making, and as adhesives for paper and board. Sodium carboxymethyl cellulose may also be used in place of gum arabic in desensitising the non-image areas of lithographic plates.

Rubber-based adhesives

Rubber-based adhesives are either applied as light brown viscous solutions in organic solvents or as less viscous lattices, usually milky-white in colour. Rubber occurs naturally as a latex, obtained by tapping the bark of rubber trees. Its chemical structure lies somewhere between that of a thermoplastic and thermosetting polymer, since its long chain molecules tend to become crosslinked due to atmospheric oxidation. The crosslinking of the molecules can be taken much further in the process of vulcanisation (page 39).

Adhesives based on natural rubber are noted for their excellent flexibility, high initial tack and a good retention of this tack over a period of several weeks. This tack retention is put to good use in making self-sealing envelopes and in other pressure sensitive applications. Rubber solutions are extremely inflammable and the latex type of adhesive is generally preferred for bonding paper and board. Rubber latex adhesives are miscible with water and usually have an alkaline pH. They provide good adhesion to a variety of surfaces including ink, varnish, lacquer, aluminium foil and plastics. Synthetic rubber adhesives have better resistance to oils, chemicals and sunlight than those based on natural rubber, although their flexibility, tack and tack retention is often inferior.

Inorganic adhesives

Sodium silicate is the only inorganic adhesive of any importance in printing and packaging. It is normally applied in aqueous solution and it sets rapidly by the loss of water. Its main use is in the manufacture of corrugated paperboard, but it may also be used for labels and for glueing aluminium foil to paper.

Synthetic resin adhesives

Synthetic resin adhesives are almost as varied as synthetic resins themselves. We have seen in an earlier chapter how molecules of simple chemical substances can be linked together by means of addition or condensation reactions to form large polymer molecules, either having a long chain or a network structure. The long chain polymers are thermoplastics capable of softening on the application of heat and of hardening again on cooling whilst the polymers with a three-dimensional lattice structure are thermosetting materials which only flow during the polymerisation process and then harden irreversibly. Another basic difference between the two types of polymers is that whilst thermoplastics are generally soluble in certain solvents, thermosetting plastics are insoluble.

Many of the various classes of synthetic polymers have been used as adhesives. Adhesives based on thermosetting resins are normally applied in a partially polymerised state, setting taking place during a curing process, which either involves heat or relies on the action of a catalyst included in the adhesive. The polymerisation of the adhesive layer in the actual joint leads to high bond strengths and these thermosetting polymers provide valuable structural adhesives. However they are only used to a limited extent in print finishing and packaging. Synthetic resin adhesives based on thermoplastic polymers are much more widely used for these purposes. They may be applied as solutions in organic solvents, as water emulsions or as hot melts. Within this group of materials, the polyvinyl resin adhesives form a particularly important class, which includes the versatile polyvinyl acetate adhesives.

Polyvinyl acetate is an addition polymer produced by the linkage of the simple monomer, vinyl acetate (Table 4.1). The presence of highly polar acetate groups give the resin excellent adhesion to many surfaces. Furthermore, its broad distribution of molecular weights means that those molecules with a relatively low molecular weight are able to contribute tack and adhesion whilst the larger polymer molecules ensure that the resin has good cohesive strength. Polyvinyl acetate emulsion often contains a small amount of polyvinyl alcohol, added to improve the emulsion stability. Plasticisers, thickening agents and solvents are among the other materials added to the various grades of polyvinyl acetate emulsions in compounding adhesives for particular applications. Although these emulsion adhesives may be diluted with water, it must be remembered that any change in their water content has a marked effect on their viscosity. Like all synthetic resin adhesives, these PVA emulsion adhesives have good resistance to attack by bacteria moulds and insects. They are used for a variety of purposes in print finishing and packaging, including adhesive binding, carton glueing, tube winding, paper cup manufacture, heatseal coatings, and cellophane or acetate window glueing.

Hot melt adhesives

Hot melt adhesives contain no water or solvents and so may be regarded as 100% solids. They are based on waxes or thermoplastic resins and their colour ranges from white to dark brown, depending on composition. Hot melt adhesives are able to provide rapid and strong bonds between a wide range of porous and non-porous substrates and they are particularly useful in laminating and bookbinding.

17. Bookbinding materials

Whilst the printer has the task of arranging a marriage between ink and paper, the binder has to blend together a much greater variety of materials in the production of a book. This is one reason why fine binding by hand remains one of the high points of craftsmanship. Today, most books are produced in what has become one of the most highly mechanised sections of the printing industry, and there is a growing need for technicians, who understand the nature and properties of the various binding materials, so that these materials can be effectively selected and controlled.

The materials commonly used in binding include:

Papers
Boards
Book covering materials: woven fabrics, non-woven materials, plastics, leathers.
Adhesives
Securing materials: thread, tape, wire.
Blocking materials: gold leaf, blocking foils.

Papers

Some of the more important classes of paper were described in Chapter 8. Most book papers have a relatively rough antique finish and a bulky open structure. Smooth glossy papers are avoided where possible because specular reflections from their surface are a distraction to the reader. The type of paper chosen for a particular book will depend on which printing process is to be used and also on whether the book is going to include halftone illustrations, line illustrations or text matter alone. An ordinary *machine finished* (MF) paper is perfectly satisfactory for a book consisting largely of text and a few line illustrations. When fine screen halftone illustrations have to be reproduced, it is necessary to use a more heavily calendered paper or even a coated paper. *Art papers* are particularly unsuitable for bookbinding and they should never be used as end papers. When they are sewn, the paper tends to tear, and if they are guarded, their coating may come away from the body of the paper. *Cartridge* papers are widely used for books printed by offset litho and also as end papers. Other papers which may be used in the actual construction of a binding include brown *kraft* papers for lining the spines of books, *banks* and *airmail* papers for guarding sections and reinforcing weak papers, *marbled* papers as sides on half-bound books and as end papers, and *tissues* for mending damaged sheets.

Boards

The three main classes of board used in bookbinding are strawboard, chipboard and millboard. Their composition and method of manufacture were outlined in an earlier chapter (page 82). The yellow coloured *strawboards* combine rigidity with low cost, but they are not particularly tough, their corners are rather easily damaged and they are prone to excessive warping. Nevertheless they are suitable for the cheaper bindings, for which permanence is of secondary importance. *Chipboards* are usually associated with carton work, but heavier grades are also widely used for the better quality bindings. Although they are based on waste paper pulps and they are less rigid than strawboards, their wearing and folding properties are much superior. Chipboards are intermediate in quality and price between the strawboards and the *millboards*, which are used for the finest quality bindings. These millboards were once made from the fibres of old tarred ropes. They were heavily calendered to give compact sheets, which were flat, hard-wearing and almost black in colour. The grey millboards in use today have similar properties and they are also based on long-fibred pulps, but these may be obtained from a variety of sources, including good quality waste paper, sacking and waste hemp.

It is important that paper and board used in making books is cut so that its machine direction runs parallel to the spine of the book. This makes for ease of opening and also helps to control warping.

Book covering materials

The covering material for a book needs to combine a pleasant appearance with strength and durability. It should be reasonably soft and pliable, and handle well without becoming too easily soiled. A wide variety of materials are available, ranging in quality and price. They fall into four main groups, namely those based on woven fabrics, non-woven materials, plastics and leather.

Woven fabrics

Woven fabrics for book covering are produced by dyeing and coating ordinary grey cloth, usually cotton. Their quality largely depends on the strength of the original weave, which may be anything from the open weave of a very light binding cloth, to the double warp closely woven material used to make buckram. The strength of one of these fabrics is closely related to the number of threads to the square centimetre. In the weaving process, the shuttle carrying the weft yarn passes through the warp as alternate threads are raised and lowered. After weaving, the fabric is desized, cleaned, bleached, washed, inspected and then cut into the required lengths. At this stage the material is normally dyed before being filled and coated with a dyed paste which is forced into the weave by heavy rollers. Finally the fabric is calendered or embossed in order to produce the desired surface finish. A filling of up to 20% is said to improve the strength of a good cloth, but a high proportion of filling pressed into a weak open cloth can lead to the filling coming away on the folds. These filling pastes are normally starch-based so the resulting fabrics are sensitive to both heat and moisture. However, by plastic coating woven cloth with solutions of various polymers, another group of book covering materials has been developed with much greater oil, water and stain resistance than ordinary bookcloths. The polymers used in making these *leather cloths* include nitrocellulose (*eg* Rexine), ethyl cellulose, polyvinyl chloride and synthetic rubbers. Solutions of these resins containing appropriate pigments are pressed into the fabric, which is then passed through a drying section where solvent is removed. These plastic coated cloths can be given attractive textured finishes by finally pressing them between heated and engraved rollers.

Non-woven materials

The wide range of non-woven book covering materials consists essentially of papers which are based on long fibred pulps and which have been given some form of surface finish. The large number of available finishes is achieved by combinations of possible surface treatments, for example, the material may be coated or uncoated, plain or embossed, printed or unprinted.

Some of these non-woven materials were primarily developed as substitutes for bookcloth. Perhaps the best known of these is Linson, which was first introduced as long ago as 1936. The basic material is produced from a pulp of the long and very strong fibres extracted from manila hemp (page 55). After the sheet has been formed on a special paper machine it is impregnated with a filler and then embossed to produce an appropriate finish, for example, simulating the characteristic appearance of book fabrics like buckram. Final processes of glazing or lamination extend the possible range of surface effects. Bookcloth substitutes of this type are able to combine good wearing properties with an attractive appearance. Ordinary printing papers and boards are also used as covering materials although obviously they are less durable than those first discussed. They are available in attractive pre-printed forms and their handling properties can be finally improved by varnishing or lamination. Plastic coated papers form another group of book covering materials. As with bookcloths, the polymer coating improves their durability and their resistance to water, oil and staining.

Plastics

In addition to the plastic coated or laminated bookcloth or paper, plastics are being increasingly used on their own as binding materials. The most popular of these sheet materials is plasticised polyvinyl chloride, PVC (page 45). Since PVC is difficult to glue to itself or to other materials, bonds are obtained by high frequency welding. The great advantage of plastics of this type is that they have excellent resistance to water, oil and most chemicals. PVC can be given a textured finish by embossing, and although plastic bound books would be regarded by many as having both an unpleasant appearance and 'feel', they are extremely functional when being used in such places as laboratories, workshops and kitchens, when more conventional covers would be quickly soiled.

Leathers

A book bound in good leather is in a class of its own both in its appearance and in its handling quality. A wide range of leathers is available for book covering. They are obtained by tanning the skins of various animals, including goats (moroccos), calves, pigs, sheep and seals. Within this range there are many different qualities available. For example, Levant Morocco is tough, hard-wearing and beautifully finished with a bold grain. Calf is much softer and less durable, but it has an attractive and delicate appearance. Pigskin is rather thick and inflexible but it is hard-wearing when well tanned. Fuller information on the various types of leathers for book covering can be found by reference to the bibliography.

Research has been carried out into the causes of deterioration of the leather bindings of old books and this has been shown to be greatly accelerated by the presence of acid. This acid may either have been left in the leather after the process of tanning or more usually the acid concentration builds up as a result of the slow absorption of sulphur dioxide gas from the polluted atmosphere of an industrial area. The conversion of this gas to sulphuric acid is catalysed by the presence of minute quantities of various metals contained in the tanned skins. Although the acid content of all leathers will build up in this way, it has been established that the fibre structure is only attacked when the tanning process or washing has removed certain water-soluble protective substances. Protection from acid attack can be restored to the leather by treating it with a dilute solution of potassium lactate. Many bookbinding leathers are now protected in this way.

An indication as to whether a particular leather is liable to decay can be obtained by carrying out an accelerated ageing test, developed by PIRA in the early years of its existence. In this test, the addition of sulphuric acid and an oxidising agent over a period of one week is equivalent to the effect of exposing the leather to an industrial atmosphere for twenty years or more.

Adhesives

The nature of adhesion and of the main classes of adhesives are discussed in the previous chapter. Bookbinding adhesives have undergone some remarkable changes in recent years, in matching the trend towards automated binding. Versions of the more traditional starch, dextrin and animal glues have remained satisfactory for many purposes, but there has been an increasing application of synthetic resin adhesives, either in the form of emulsions or as hot melts. These synthetic adhesives have had to be developed to meet the problems of adhesion posed both by new covering materials and by the extremely high speeds of production. Adhesive binding lines are now capable of turning out 10 000 books an hour.

Hot melt adhesives are extensively used by the large producers of paperback books. In the 'one-shot' system a hot melt binds the pages, forms the spine and holds the cover which is laid on the adhesive layer before it cools. In the 'two-shot' system an emulsion adhesive is first used to penetrate and bind the pages, and then the application of a hot melt strengthens the spine and glues the cover. Another approach to automated binding is made in the use of a specially formulated polyvinyl acetate emulsion adhesive used in conjunction with microwave drying.

Securing materials

Thread

Strong and durable *thread* forms a small but important component in a sewn book. Traditionally all bookbinding thread was manufactured from cotton or linen but more recently synthetic fibres like nylon and terylene have been used and blends of these synthetic fibres with natural fibres have become popular.

Unbleached *linen* thread, which is stronger and more durable than cotton, is normally used for hand sewing. It consists of several threads twisted together and is sold in hanks in various thicknesses. The thickness of a linen thread is defined by a number; those used in binding normally range from 16 to 25; the larger the number the thinner the thread.

Cotton thread is always associated with machine sewing. Like linen, it is manufactured in a range of thicknesses given by a number. The various grades are designed for use on the many different types of sewing machines. The choice of thickness should be governed by the thickness of the paper and by the number of sections. Cotton which is either too thin or too thick can make it impossible to obtain the right amount of 'swell' needed to form a well shaped book.

Cotton may play several different roles in the binding of a book. Apart from its use in making bookcloth and as a sewing thread, it appears as the closely woven tape needed for sewing the sections of some books and as the open-weave fabric known as *mull*, used to reinforce the glued spine.

Although *nylon* thread is extremely strong and relatively cheap, its main drawback in bookbinding is its high degree of elasticity. After cutting, this elasticity of nylon causes it to contract and pull back, so loosening the end sections. *Terylene* is better in this respect and has excellent strength. By blending synthetic fibres like terylene with the natural fibres like cotton, the best properties of each fibre can be combined in a single thread. These blended threads are becoming increasingly popular in bookbinding.

Wire

Stitching wire is supplied either in reels or coils in grades ranging from the very thin 30 gauge wire (diameter 0·315mm) to the much thicker 18 gauge (diameter 1·220mm) used on heavy duty machines. Most of the wire used in the printing industry is galvanised mild steel, but where extra strength is needed to penetrate thick piles of hard paper, steels with a higher carbon content may be used. Among other metals that can be used, copper has the advantage that it will not rust, but its softness leads to frequent stoppages on stitching machines. However, copper-coated steel wires combine the advantages of the two metals. Where the appearance of the stitching wire is important, attractive materials like stainless steel, brass and related alloys (*eg* Alumenoid), may be used, but all these metals are very much more expensive than normal stitching wire. Coloured wires have become popular in recent years, particularly for spiral binding. These generally consist of mild steel coated with a pigmented lacquer or paint.

Blocking materials

Gold leaf

Gold has been used in the decoration of books for many centuries. Quite apart from its brilliant appearance and excellent resistance to tarnishing, it is the most malleable of metals and it can be beaten flat into a leaf, only 100nm thick. Gold decoration was originally applied by cutting the leaf by hand and laying it on the article, which had been previously sized with shellac or egg albumen. Later the blocking operation was simplified by carrying the gold leaf on the surface of a roll of waxed paper, and in a further development a coating of shellac was laid over the gold leaf so that it became possible to carry out continuous blocking on a hot press stamping machine.

Blocking foils

These developments with gold leaf were some of the first steps in the evolution of one of the oldest crafts into a modern industry. More recently the range of blocking foils has been greatly extended by the use of less expensive materials including aluminium, alloys like Dutch metal containing copper and zinc, and pigmented coatings. The tedious process of beating gold down to leaf thickness has been replaced by the vacuum deposition of the metal on to the base material. Paper, the original carrier for gold, has been replaced by cellulose film, cellulose acetate and most recently by polyester films like 'Melinex' (page 46).

However, like the paper-wax-gold-shellac material mentioned earlier, blocking foils still consist of four essential components, namely:

(1) a base material, eg a polyester film;
(2) a layer of a special wax which will melt under the influence of the heat of the embossing die;
(3) a layer of metal or pigmented coating; and
(4) a layer of adhesive chosen to match the material being decorated, eg a foil being blocked on to polyvinyl chloride (PVC) must carry a special adhesive coating.

The total thickness of the layers laid on to the base film varies between one and two micrometres.

Appendix A

SOME REFERENCES FOR FURTHER READING

Chapter 2, Metals for platemaking

CARTWRIGHT, H M, *Ilford Graphic Arts Manual*, Vols 1 and 2, Ilford Ltd

COUPE, R R, *Science of Printing Technology* (Chapters on Surface Chemistry and Electrochemistry), Cassell

MARSHALL, G R, *Introduction to Science for Printers*, Heinemann

PATEMAN, F, and YOUNG, L C, *Printing Science* (Chapter on Metals for Platemaking), Pitman

Chapter 3, Printing alloys

Printing Metals, Fry's Metal Foundries Ltd

Printing Metals, Pass Printing Metals Ltd

PATEMAN, F, and YOUNG, L C, *Printing Science* (Chapter on Metals for Casting), Pitman

MARSHALL, G R, *Introduction to Science for Printers*, Heinemann

WEAVER, F D, *The Lead-Antinomy-Tin System*, Journal of the Institute of Metals Vol. LVI, No. 1, 1935.

Chapter 4, Polymers

BRYDSON, J A, *Plastics Materials*, Iliffe Books

MILES, D C, and BRISTON, J H, *Polymer Technology*, Temple Press

SANDERS, K J, *Identification of Plastics and Rubbers*, Chapman & Hall

ASKEW, F A, editor, *Printing Ink Manual*, published for the Society of British Printing Ink Manufacturers by Heffer, 2nd edition 1969

GAIT, A J, and HANCOCK, E G, *Plastics and Synthetic Rubbers*, Pergamon

LEVER, A E, and RHYS, J A, *The Properties and Testing of Plastics Materials*, Temple Press

PATEMAN, F, and YOUNG, L C, *Printing Science* (Chapter on Polymers and Printing), Pitman

Appendix A

PROCEEDINGS OF 1966 DUPLICATE PLATE CONFERENCE, London, Duplicate
Plate Research and Technical Committee
YOUNG, L C, 'Polyethylene and Polypropylene', *British Printer*, Vol. LXXX,
No. 3, March, 1967
YOUNG, L C, 'Polyesters', *British Printer*, Vol. LXXX, No. 5, May, 1967
BOLL, O L, *Plastics and Rubbers in Printing*, Plastotype Ltd
YOUNG, L C, 'Thermosetting Plastics', *British Printer*, Vol. LXXX, No. 7,
July, 1967

Chapters 5–9, Paper and board

General

BRITISH PAPER AND BOARD MAKERS ASSOCIATION, *Papermaking*. Technical
Section, 1965
GILMOUR, S C, editor, *Paper, Its Making, Merchanting and Usage*, National
Association of Paper Merchants & Longmans Green
NORRIS, F H, *The Nature of Paper and Board*, Pitman
HIGHHAM, R R A, *Handbook of Paper Making*, Business Books, 1968
HIGHHAM, R R A, *Handbook of Paperboard and Board*, Business Books, 1970
GRANT, J, *Laboratory Handbook of Pulp and Paper Manufacture*, Arnold
MACDONALD, R G, editor, *Papermaking and Paperboard Making*, McGraw Hill,
1970
NADELMAN, A N, and BALDAUF, G H, *Coating, Formulations, Principles and
Practices*, Lockwood Trade Journals
BRYANT, P J, 'Paper Coating Methods', *Printing Technology*, Vol. 12, No. 1,
April, 1968
BLOKHUIS, G, 'Paper for Photogravure Printing', *Printing Technology*, Vol. 10,
No. 1, April, 1966
PIRA Buying Guides, No. 1 *Letterpress Papers and Boards*, No. 2 *Lithographic
Papers and Boards*
REED, R F, *What the Printer should know about Paper*, Graphic Arts Technical
Foundation, 1970

Properties and testing methods

COUPE, R R, 'Paper and Ink Testing', *Paper and Ink*, Proceedings of PATRA
Letterpress Conference, October, 1955
CLARKE, B, 'The Microscope in Paper Research', *Printing Technology*, Vol. IV,
No. 2, December, 1960
ADAMS, J M, *Optical Measurements in the Printing Industry*, Pergamon
PRITCHARD, E J, and CAIRNS, J A, *Air Conditioning for Printers*, PATRA, 1962

CHAMBERLIN, G J, and COUPE, R R, 'Testing the Surface pH of Papers', *Printing Technology*, Vol. VI, No. 2, November, 1962

FRANKLIN, A T, and WILSON, C M, 'The Dimensional Stability of Paper—A Survey', *Printing Technology*, Vol. VII, No. 1, July, 1963

PARKER, J R, 'An Air Leak Instrument to measure the Printing Roughness of Paper and Board', *Paper Technology*, Vol. 6, No. 2, April, 1965

REED, R F, and STANTON, D R, *Instruments and Controls for the Graphic Arts Industries*, Graphic Arts Technical Foundation, 1971

PARKER, J R, 'The Use of Instruments to predict the Printability of Paper', *Paper Technology*, Vol. 7, No. 6, pp. 560–564, December, 1966

DAVEY, B H, 'Surface Smoothness', *British Printer*, Vol. LXXX, No. 12, December, 1967

DAVEY, B H, 'Measuring Absorbency', *British Printer*, Vol. LXXI, No. 8, August, 1968

LEE-FRAMPTON, J, and ARMITAGE, A, 'Measurement of Colour and Brightness', *The Paper Maker*, March, 1969

SWAN, A, 'Realistic Paper Tests for Various Printing Processes', *Printing Technology*, Vol. 13, No. 1, April, 1969

IGT INFORMATION LEAFLETS, *Application of the IGT Printability Testers A1, A2 and AC2*

FRANKLIN, A T, 'Print Quality and Runnability of Coated Paper', *Printing Technology*, Vol. 14, No. 1, April, 1970

MUNDAY, F D, 'Relationship between Paper Properties and Print Quality', *Printing Technology*, Vol. 14, No. 1, April, 1970

PARKER, J R, 'Measurement and Control of linting from Web-Offset Newsprint', *Printing Technology*, Vol. 14, No. 1, April, 1970

BLUNDEN, B and PEACOCK, E W, 'Paper Properties that Count', *British Printer*, Vol. 83, No. 11, Nov, 1970

Chapters 10–14, Printing ink

General

ASKEW, F A, editor, *Printing Ink Manual*, Published for the Society of British Printing Ink Manufacturers by Heffer, 2nd edition, 1969

LARSEN, L M, *Industrial Printing Inks*, Reinhold, 1962

APPS, E A, *Ink Technology for Printers and Students*, Leonard Hill, 1963

PARKER, D H, *Principles of Surface Coating Technology*, Interscience, 1965

CHATFIELD, H W, editor, *Science of Surface Coatings*, Benn, 1962

OIL AND COLOUR CHEMISTS ASSOCIATION, *Paint Technology Manuals*, Chapman and Hall

Appendix A

MORGANS, W H, *Outlines of Paint Technology*, Griffin, 1969
WATFORD COLLEGE OF TECHNOLOGY, *Pigment Dispersion Bibliography 1960–70*

Properties and testing methods

PEACOCK, S, 'Some practical aspects of Ink Performance', *British Ink Maker*, Vol. 9, No. 4, August, 1967

PATEMAN, F, 'Fluid Flow', *British Printer*, Vol. LXXX, No. 8, August, 1967

CARTWRIGHT, P F S, 'Specifying the Rheological Properties of Lithographic Printing Inks', *Printing Technology*, Vol. VIII, No. 2, December, 1964

WHITFIELD, G W, 'Rheology of Printing Ink (Letterpress and Litho)', *British Ink Maker*, Vol. 7, No. 2, February, 1965

HARE, D G, and TURNER, T A, 'Rheological Properties of Offset Printing Inks in relation to Pigmentation', *British Ink Maker*, Vol. 12, No. 3, May, 1970

DOUGLAS, A F, and SPAULL, A J B, 'Rheology of Printing Ink. Tentative Explanation of the Role of Visco-elasticity', *British Ink Maker*, Vol. 12, No. 1, November, 1969

CARTWRIGHT, P F S, 'Effect of Water on the Rheological Properties of Lithographic Tin-printing Inks', *British Ink Maker*, Vol. 8, No. 4, August, 1966

CARTWRIGHT, P F S, 'Measurement of the Thixotropy of Lithographic Printing Inks', *British Ink Maker*, Vol. 8, No. 2, February, 1966

PATEMAN, F, 'Viscometers', *British Printer*, Vol. LXXX, No. 10, October, 1967

HANSEN, W, 'A Comparison of Two Methods for Calculating Viscosity based on Measurements on a Falling Bar Viscometer', *British Ink Maker*, Vol. 11, No. 4, August, 1969

WHITE, I C, and DANIELS, B, *Improvements in the Use of Falling Rod Viscometers*, TAGA proceedings, 1970, p. 11

COUPE, R R, 'Tack in Printing Inks', *Litho Printer*, June, 1965

PATEMAN, F, 'The Measurement of Tack', *British Printer*, Vol. LXXXI, No. 12, December, 1968

CARTWRIGHT, P F S, 'The Churchill Tackmeter', *British Ink Maker*, No. 4, 224, 1962

VAN HESSEN, B, 'The Tack-O-Scope', *American Ink Maker*, 41, No. 1, 22, 1963

SCHAEFFER and CHAKRAVARTI, 'The NPIRI-Chatillon Tack Finger', *American Ink Maker*, No. 10, Vol. 47, October, 1969

BASSERNIR, R, 'Testing the NPIRI-Chatillon Tack Finger', *American Ink Maker*, No. 11, Vol. 48, November, 1970

CAHIERRE, L, 'A Simple Apparatus for measuring the Tack of Printing Inks', *Advances in Printing Science and Technology*, Vol. 5

SOUTH, G R, 'A Comparison of Tackmeters', *British Ink Maker*, Vol. 10, No. 3, May, 1968

CATFORD, J R, 'Instrumentation in the Printing Industry', *Instrument Review*, October and November, 1965

PATEMAN, F, 'Densitometers', *British Printer*, Vol. LXXX, No. 4, April, 1967

ADAMS, J M, *Optical Measurements in the Printing Industry*, Pergamon

WHITFIELD, G W, 'The Employment of Scientific Instruments in Ink Making', *British Ink Maker*, Vol. 10, No. 1, November, 1967

REED, R F, and STANTON, D R, *Instruments and Controls for the Graphic Arts Industries*, Graphic Arts Technical Foundation (New York)

'The PATRA Ink Film Thickness Gauge', *British Ink Maker*, Vol. 7, No. 4, August, 1965, and Vol. 8, No. 3, May, 1966

BISSET, D E, and WOODS, D W, 'The Practice and Interpretation of Certain Ink Testing Methods', *Printing Technology*, Vol. 14, No. 3, December, 1970

POND, K, and RICHARDSON, P G, 'Printability Measurements', *Printing Technology*, Vol. V, No. 1, August, 1961

CARTWRIGHT, P F S, SMITH, C H, and CARR, W, 'Pigment, Ink and Printing', *Printing Technology*, Vol. 11, No. 2, July, 1967

FRANKLIN, A T, 'Exploration of Some Ink/Paper Interactions', *British Ink Maker*, Vol. 12, No. 3, May, 1970

BIRKETT, P, 'Lightfastness', *British Printer*, Vol. LXXXI, No. 11, 1968

HOBBS, H, 'Lightfastness of Prints', *British Ink Maker*, Vol. 12, No. 1, November, 1969

Chapter 15, Light-sensitive materials

BAINES, H, and BOMBACK, E S, *The Science of Photography*, Fountain Press

WALLS, H J, *How Photography Works*, Focal Press

LOBEL, L, and DUBOIS, M, *Basic Sensitometry*, Focal Press

GOROKHOVSKII, YU N, and LEVENBERG, T M, *General Sensitometry*, Focal Press

MEES, C E K, and JAMES, T H, *Theory of the Photographic Process*, Macmillan

JAMES, T H, and HIGGINS, G C, *Fundamentals of Photographic Theory*, Morgan and Morgan

L P Clerc's Photography: *Theory and Practice*, Edited D A Spencer, Focal Press

The Manual of Photography (formerly *The Ilford Manual of Photography*) Edited Alan Horder, Focal Press

NEBLETTE, C B, *Fundamentals of Photography*, Van Nostrand Reinhold, 1970

KOSSAR, J, *Light Sensitive Systems*, Wiley

SCHAFFERT, R M, *Electrophotography*, Focal Press

YOUNG, L C, 'Dichromated Colloids', *British Printer*, Vol. LXXX, No. 3, March, 1968

YOUNG, L C, 'Diazo Compunds (Introduction)', *British Printer*, Vol. LXXX, No. 11, November, 1967

DINNABURG, M S, *Photosensitive Diazo Compounds*, Focal Press

CHAMBERS, E, *Camera and Process Work*, Benn

CHAMBERS, E, *Photolitho Offset*, Benn

CARTWRIGHT and MACKAY, *Rotogravure*, MacKay Publishing Corporation (New York)

YOUNG, L C, 'Photopolymers', *British Printer*, Vol. LXXXI, No. 1, January, 1968

Chapter 16, Adhesives

ALNER, D J, editor, *Aspects of Adhesion 1 and 2*, University of London Press

PARKER, R S R, and TAYLOR, P, *Adhesion and Adhesives*, Pergamon

ADHESIVES DIRECTORY, O'Connor, A S, Richmond, Surrey

MCGUIRE, E P, *Packaging Adhesives*, Palmerton Publishing, New York

OPIE, W J, and HALIFAX, J B, 'Adhesion and Adhesives in the Graphic Industries', *Printing Technology*, Vol. 7, No. 1, July, 1963

'Adhesives for the Printer', *British Printer*, Vol. LXXXI, No. 8, August, 1968

Chapter 17, Binding materials

CLOUGH, E A, *Bookbinding for Librarians*, Association of Assistant Librarians

UPTON, P G, 'A Technologist looks at Books', *Printing Technology*, Vol. III, No. 2, January, 1960

DAY, F T, 'Surface Finish and Finishing Processes on Paper and Board', *British Printer*, Vol. LXXVIII, No. 8, August, 1965

KAHLMANN, G, 'All That Glitters . . .', *British Printer*, Vol. LXXXIII, No.7, July, 1970

'The Linson Story', *British Printer*, Vol. LXXVIII, No. 8, August, 1965

Production Guides on Metallic Foils, George Whiley Ltd, Victoria Road, Ruislip, Middlesex

CHAVE, L, 'Survey of Book Covering Materials', *Print in Britain*, November, 1967

READING, C L, 'Wire Stitching Machines', *British Printer*, Vol. LXXVII, No. 9, September, 1964

Appendix B

SELECTED LIST OF TESTING EQUIPMENT FOR PAPER, INK AND OTHER PRINTING MATERIALS WITH NAMES AND ADDRESSES OF SUPPLIERS

Note: Many of the instruments listed below may be obtained through distributors specialising in testing equipment for the paper and printing industries, notably H. E. Messmer Ltd. in the United Kingdom and Testing Machines Inc. in the United States.

(a) Instruments for testing paper	Manufacturer or Distributor
Grammage (Basic weight or substance)	
Paper scales	Amthor
	E J Cady
	Lhomargy
	Lorentzen & Wettres
	H E Messmer
	Testing Machines
	Toledo Scale
Thickness	
Deadweight spring operated and hand-held micrometers	B C Ames
	Amthor
	J E Baty
	E J Cady
	Custom Scientific
	Lorentzen & Wettres
	Lhomargy
	H E Messmer
	Testing Machines

Appendix B

	Manufacturer or Distributor
Mechanical properties	
Tensile strength testers	Amthor
	Instron
	Karl Frank
	Lhomargy
	Lorentzen & Wettres
	H E Messmer
	Tensometer
	Testing Machines
	Thwing-Albert
	Van der Korput
Tearing strength testers	Karl Frank
	Liberty Engineering
	Lorentzen & Wettres
	H E Messmer
	Testing Machines
	Thwing-Albert
Bursting strength testers	Amthor
	E J Cady
	Karl Frank
	Lhomargy
	Lorentzen & Wettres
	H E Messmer
	B F Perkins
	Testing Machines
	Zellstoff und Papierfabrik
Folding endurance testers	Karl Frank
	Lhomargy
	Lorentzen & Wettres
	H E Messmer
	Testing Machines
	Tinius Olsen
Stiffness testers	Karl Frank
	Gardner Laboratory
	H A Gaydon
	W & L E Gurley
	Lhomargy
	Lorentzen & Wettres

Manufacturer or Distributor

H E Messmer
Taber Instruments
Testing Machines
Thwing-Albert

Carton creasing properties

Carton board creaser (PIRA) Headland Works
                             Testing Machines
Carton crease gauge (PIRA)   H W Wallace
                             Testing Machines
Concora boxboard scorer   Liberty Engineering
                             H E Messmer
                             Testing Machines
Crease stiffness tester (PIRA) C F Taylor
                             Testing Machines
Creasing tester   Karl Frank
                             H E Messmer

Moisture content

Moisture meters   Hart Moisture Meters
                             Marconi
                             Moisture register
Infra tester   Promin
Dynatronic IR moisture analyser S Garcia
Moisture content balance   William Webb

Temperature

PIRA stack thermometer   British Rototherm

Relative Humidity

Humidity indicators and recorders Cambridge Instruments
                             C F Casella
Cambridge paper hygroscope   Cambridge Instruments
Casella sword hygroscope   C F Casella
Hygro champ   Hygrodynamics

	Manufacturer or Distributor
Prüfbau sword probe	Dr Ing Dürner
	Rodney O Berger
Saber champ	Hygrodynamics
Spatulate blades	Hart Moisture Meters
Sword sensor	Testing Machines
Turbojet sword hygrometer	Turbojet control

Dimensional stability

Clark hygroexpansimeter	Thwing-Albert
L & W expansion tester	Lorentzen & Wettres
	H E Messmer
Neenah expansimeter	Martin Sweets
	H E Messmer
	Testing Machines
PIRA expansion measuring	H W Wallace
apparatus	H E Messmer
	Testing Machines
TMI expansion and shrinkage tester	Testing Machines
	H E Messmer

Surface properties

Fluffing tendency

PIRA fluff tester	H W Wallace
	Testing Machines
IGT loose paper particle tester	Rudolf Meyer's
	H E Messmer
	K E Saunders

Pick resistance

Dennison paper testing waxes	Dennison Manufacturing
	Testing Machines
GATF pick tester	Stouffer Graphic Arts
GLF picking tester	Wennberg Apparater AB
Scott internal bond tester	Scott Testers
Printability testers	see below

Manufacturer or Distributor

Instruments for the making of prints and the assessment of printability

Domtar printograph	Testing Machines International of Canada
	Testing Machines
Film applicators	Sheen Instruments
Fogra printability tester	Karl Frank
	H E Messmer
Gravure proofing press	Winstones
GRI gravure printability tester	Huck
IGT printability tester	Rudolf Meyer's
	H E Messmer
	Testing Machines
K gravure proofer	RK Chemical
K flexo proofer	RK Chemical
K film proofer	RK Chemical
Minipress	Andersson and Sørensen
	H E Messmer
	Testing Machines
Precision printing gage	Precision Gage & Tool
Proof print tester	Duncan Lynch
Prüfbau test printer	Dr Ing Dürner
	Rodney O Berger
'Quickpeek' colour proofing kit	Thwing-Albert
'REL' adjustable film spreader	Research Equipment

Smoothness

Bekk smoothness and permeability tester	Van der Korput
	H E Messmer
	Testing Machines
Bendtsen smoothness and porosity tester	Andersson and Sørensen
	H E Messmer
	Robbins Instrument
	Testing Machines
FOGRA smoothness tester	Dr Ing Dürner
	Rodney O. Berger
Gurley-Hill smoothness-porosity-softness tester	W & L E Gurley
	H E Messmer
	Testing Machines

Appendix B

	Manufacturer or Distributor
Microcontour ink	Lorilleux & Bolton
Parker print-surf roughness tester	H E Messmer
	Testing Machines
PIRA smoothness tester	British Industrial Tooling
Sheffield smoothness gauge	Bendix Corporation
Talysurf	Rank Precision Industries
Printability testers	see 'Instruments for making prints'

Oil absorbency

IGT penetration volumeter	Rudolf Meyer's
	H E Messmer
	K E Saunders
	Testing Machines
Lhomargy penetration measurer	Lhomargy
PIRA surface oil absorption tester	H W Wallace
	Testing Machines
Vanceometer	Hillside Laboratories
	Testing Machines

Ink absorbency

Donnelly-Hull wipe test inks	Inmont
K & N testing ink	K & N Laboratories
	Fishburn
K & N absorption tester	Wennberg Apparater AB
	Testing Machines
Printability testers	see 'Instruments for making prints'

Water absorbency

Cobb sizing tester	Kent Engineering
Gurley-Cobb sizing testers	W & L E Gurley
	H E Messmer
	Testing Machines
Galvanic sizing testers	H E Messmer
	Testing Machines
Sizing tester	Lorentzen & Wettres

	Manufacturer or Distributor
Porosity	
Air permeability testers	Karl Frank
	Lorentzen & Wettres
	H E Messmer
	Testing Machines
Densometers	W & L E Gurley
	Lorentzen & Wettres
	H E Messmer
	Testing Machines
Permeometers	W & L E Gurley
	H E Messmer
	Testing Machines
	see also under 'Smoothness testers'
pH	
Indicators and indicator papers	British Drug Houses
	Paul Frank
	Micro-Essential Laboratory
Colour comparators	Tintometer
PIRA-Lovibond colour slides	Tintometer
pH meters	Suppliers include:
	Analytical Instruments
	Beckmann Instruments
	Cambridge Instruments
	Corning Glass Works
	Electronic Instruments
	W G Pye

(b) Instruments for testing printing ink

Fineness of grind gauges	Precision Gage & Tool
	Research Equipment
	Sheen Instruments
	Testing Machines

Appendix B

	Manufacturer or Distributor
Viscometers	
Efflux (flow cups)	Gallenkamp
	General Electric
	Griffin & George
	Norcross
	Research Equipment
	Sheen Instruments
	Stanhope-Seta
	Testing Machines
	Townson & Mercer
Laray type	Barker & Aspey
	Churchill Instruments
	Jacobsen
	Lhomargy
	Testing Machines
Rotational	Brookfield Engineering
	Churchill Instruments
	Contraves
	Ferranti
	Fisher Scientific
	Polyscience Corporation
	Sangamo Controls
	Sheen Instruments
	Testing Machines
	Viscosel Corporation
Cone and plate	Contraves
	Ferranti
	Martin Sweets
	Research Equipment
	Sangamo Controls
Tack	
Churchill-Metal Box tackmeter	Churchill Instruments
GATF inkometer	Thwing-Albert
	H E Messmer
	Testing Machines
Graded tack pastes	Mander Kidd
Laray tackmeter	Diaf A/S
	Barker & Aspey

	Manufacturer or Distributor
NPIRI-Chatillon tack finger	Chatillon
	International Engineering
PIRA tackmeter	H A Gaydon
	Testing Machines
Tack-O-Scope	Rudolf Meyer's
	K E Saunders

Press stability

BK drying recorder	Mickle Laboratory
Gardner drying time recorder	Gardner Laboratory
PIRA ink drying time recorder	PIRA

Drying time on paper

IGT drying time recorder	Rudolf Meyer's
	H E Messmer
	K E Saunders
Laray drying meter	Lhomargy
NPIRI drying time recorder	NPIRI
PIRA print drying time recorder	PIRA

Lithographic performance

Pope and Grey emulsification tester	Thwing-Albert
	H E Messmer

Ink film thickness

Cheville wet film thickness gauge	Gardner Laboratory
Churchill-PIRA ink monitor (measurements on ink roller on press)	Churchill Instruments
Interchemical wet film thickness gauge	Gardner Laboratory
	Wentworth Instruments
Sheen wet film thickness gauge	Sheen Instruments

Rub resistance

Adams wet rub tester	H E Messmer
	Testing machines

	Manufacturer or Distributor
Bendtsen rub-off tester	Andersson and Sørensen
	H E Messmer
	Robbins Instrument
	Testing Machines
PIRA rubproof tester	H W Wallace
'REL' abrasion test apparatus	Research Equipment
Sutherland rub tester	Brown Co
	Testing Machines
Taber abraser	Taber Instruments
	Gardner Laboratory
	H E Messmer
	Testing Machines

Lightfastness

Fade-ometer	Atlas Electric Devices
	Wentworth Instruments
Weather-ometer	Atlas Electric Devices
	Wentworth Instruments
Xenotest	John Godrich

(c) Instruments for meauring the optical properties of paper and print

Gloss

EEL gloss meter	Evans Electroselenium
Gardner gloss meter	Gardner Laboratories
	Testing Machines
	Wentworth Instruments
Hunterlab multipurpose gloss meter	Hunter Associates
Photovolt gloss meter	Photovolt Corporation
	Testing Machines
Sheen gloss meters	Sheen Instruments
Zeiss GP2 Goniophotometer	Carl Zeiss

Opacity

B & L opacimeter	Bausch & Lomb
Eddy opacity tester	Thwing-Albert
	Testing Machines

	Manufacturer or Distributor
EEL opacimeter	H E Messmer
Elrepho	Carl Zeiss
	Degenhardt
Opacitor	Paper Facts and Figures
Opacity guide (standard papers)	PIRA
Sheen opacity reflectometer	Sheen Instruments

Reflectometers and Densitometers

Densichron	Sargent-Welch
	Phillips Engineering
Densilux	Dahica Instruments
	Paul Murer
Densitocolor	Karl Heitz
Densitometer DS DM-270	Dainippon Screen Mfg.
EEL densitometer	Evans Electroselenium
EEL reflectometer	Evans Electroselenium
GAM digital densitometer	Graphic Arts Manufacturing
Gretag densitometer	Gretag
	Crosfield
Ink spotter	Howson-Algraphy
	(Northern)
Lovibond tintometer	Tintometer
SD 602 Luxometer	Electronic Mechanical
Photolog	Howson-Algraphy
	(Southern)
Photovolt	Photovolt Corporation
	H E Messmer
Quantalog densitometer	Macbeth Corporation
	(Kollmorgen)
Quantascan	Quantametric Devices
Soniscop densitometer	Soniscop Electronique
Tri-Colour-Phot	Phototronic
Volomat densitometer	Photounion
	Industrial Photo
	Equipment

(a) Colour Measuring Instruments

The following companies are suppliers of colour measuring instruments

Applied Research Laboratories

Appendix B

Automatic Control Devices
Bausch and Lomb
Beckman Instruments
Davidson and Hemmendinger
Diano Corporation
Gardner Laboratory
W Harrison
Hunter Associates Laboratory
Kollmorgen Corporation
 (Colour Systems Division)
Livingstone Electronic
Joyce Loebl
Martin Sweets
Metallurgical Services
 Laboratories
Neotec Instruments
Optica UK Ltd
Pretema
Teledyne
Tokyo Shivaura Electric
Unicam Instruments
Carl Zeiss Instruments

(e) **Instruments for testing adhesives**	**Manufacturer or Distributor**
Concora folder gluer	Liberty Engineering Testing Machines
PCA glueability tester	Testing Machines
Perkins-Weyco glueability tester	B F Perkins
PIRA adhesive setting time tester	H W Wallace
PIRA book page puller	H W Wallace
Thwing-Albert TAC tester	Thwing-Albert Testing Machines
Werle tack tester	Thwing-Albert Testing Machines

(f) **Other instruments used for testing printing materials**

PIRA type alignment projector   Newbold & Bulford

	Manufacturer or Distributor
Rubber hardness testers	Lhomargy
	Rex Gauge
	Shore Instrument
	H W Wallace
Paper roll hardness tester	Testing Machines
	Electronic Automation
Electric incinerator (ash determination)	H E Messmer
Web tension meters	Testing Machines
Baldwin-Dunlop Statigun	Nuclear Enterprises
(measurement of electrostatic charge)	

Note: Although every effort has been made to include those intruments which are commonly used for testing printing materials in the United Kingdom and in the United States, it is not claimed that the above list is comprehensive.

NAMES AND ADDRESSES OF SUPPLIERS OF TEST EQUIPMENT FOR PRINTING MATERIALS

B C Ames Co, Waltham, Mass, USA

Amthor Testing Instrument Co. Inc, 45–53 Van Sinderen Avenue, Brooklyn, NY, USA

Analytical Measurements Inc, 31 Willow St, Chatham, New Jersey, USA

Andersson and Sørenson, Niels Juelsgade 9–11, Copenhagen, Denmark

Applied Research Laboratories Ltd, Wingate Road, Luton, Beds, UK

Atlas Electric Devices Co, 4114 North Ravenswood Avenue, Chicago, Illinois, USA

Automatic Control Devices Inc, Box 244, Bethel, Connecticut, USA

Baird & Tatlock Ltd, Freshwater Road, Chadwell Heath, Essex, UK

Barker & Aspey Ltd, 144 Leeds Road, Hull, Yorks, UK

J E Baty & Co Ltd, Burgess Hill, Sussex, UK

Bausch & Lomb Inc, 635 St Paul Street, Rochester 2, NY, USA

Beckman Instruments Inc, 2500 Harbour Boulevard, Fullerton, California, USA

Bendix Corporation, Automation and Measurement Division, Box 893, Dayton, Ohio, USA

Rodney O Berger, 3222 Rolston St, Fort Wayne, Indiana, USA

British Drug Houses, Poole, Dorset, UK

British Industrial Tooling (London) Ltd, Doherty Works, Kingston Bridge, Kingston-upon-Thames, Surrey, UK

British Rototherm Co Ltd, Kenfig Industrial Estate, Nr Port Talbot,
 Glamorgan, South Wales, UK
Brookfield Engineering Laboratories Inc, 240 Cushing St, Stoughton,
 Mass, USA
Brown Co, Kalamazoo, Michigan, USA

E J Cady & Co, 1915 N Harlem Avenue, Chicago, Illinois, USA
Cambridge Instruments Co Ltd, Friern Park, North Finchley, London N12,
 UK
Cambridge Instruments Co Inc, 73 Spring St, Ossining, NY, USA
C F Casella & Co Ltd, Regent House, Britannia Walk, London N17, UK
John Chatillon & Sons, 83–30 Kew Gardens Rd, Kew Gardens, NY, USA
Churchill Instrument Co Ltd, Walmgate Road, Perivale, Greenford,
 Middlesex, UK
Control and Instrumentation Ltd, 21 Foxley Lane, Purley, Croydon, Surrey,
 UK
Contraves Industrial Products Ltd, Times House, Station Approach,
 Ruislip, Middlesex, UK
Corning Glass Works, 80 Houghton Park, Corning, NY, USA
Crosfield Electronics Ltd, 766 Holloway Road, London N19
Custom Scientific Instruments Inc, 13 Wing Drive, Whippany, New Jersey,
 USA

Dainippon Screen Manufacturing Co Ltd, Horikawa, Kuramaguchi,
 Kamikyo-ku, Kyoto, Japan
Davidson & Hemmendinger, 46 London Road, Reading, Berks, UK
A H Degenlardt & Co Ltd, 20–22 Mortimer Street, London W1, UK
Dennison Manufacturing Co Ltd, Colonial Way, Watford, Herts, UK
Dennison Manufacturing Co, Framingham, Mass, USA
Diaf A/S, 25–29 Vibevej, DK 2400, Copenhagen NV, Denmark
Diano Corporation, PO Box 231, 123 Central St, Foxboro, Mass, USA
Dr-Ing Herbert Dürner, 8123 Peilbenberg, Munchen, West Germany
Duncan-Lynch Precision Tools Ltd, Oak Tree Works, Molly Millars Lane,
 Wokingham, Berk, UK

Electronic Automation System Inc, 2957 Alt Boulevard, Grand Island,
 NY, USA
Electronic Instruments Ltd, Hanworth Lane, Chertsey, Surrey, UK
Electronic Mechanical Products Co, 929–935 Atlantic Avenue, Atlantic City,
 New Jersey, USA
Evans Electroselenium Ltd, St Andrew's Works, Halstead, Essex, UK

Ferranti Ltd, Moston, Manchester M10, UK
Ferranti Electric Inc, East Bethpage Rd, Plainsview, NY, USA
Fishburn Printing Ink Co Ltd, 94 St Albans Road, Watford, Herts, UK
Fisher Scientific Co, 711 Forbes Avenue, Pittsburgh, Pa, USA
Karl Frank GMBH, 68 Mannheim-Rheinau, West Germany
Paul Frank Division of Fil-Chem Inc, 153 E 26th St, New York, NY, USA

Garcia Ltd, 780 Seven Sisters Road, London N15, UK
Gardner Laboratory Inc, PO Box 5, East Syracuse, NY, USA
H A Gaydon & Co Ltd, 93 Lansdown Rd, Croydon, Surrey, UK
General Electric Co, Instrument Dept, Nela Park, Cleveland, Ohio, USA
John Godrich, Ludford Mill, Ludlow, Shropshire, UK
Graphic Arts Manufacturing Co, 2518 South Boulevard, Houston, Texas,
    USA
Graphic Arts Technical Foundation Inc, 4615 Forbes Avenue, Pittsburgh,
    Pa, USA
Gravure Technical Association Inc, 60E 42nd St, New York, NY, USA
Griffin and George Ltd, Ealing Rd, Alperton, Wembley, Middlesex, UK
W & L E Gurley, 514 Fulton St, Troy, NY, USA

W Harrison, 74 Liverpool Old Rd, Much Hoole, Preston, Lancs, UK
Hart-Moisture-Meters Inc, 336 West Islip Boulevard, West Islip, Long
    Island, NY, USA
Headland Works Ltd, 10 Melon Rd, Peckham, London SE15, UK
Karl Heitz Inc, 979 Third Avenue, New York, NY, USA
Hilger-IRD Ltd, 98 St Pancras Way, Camden Road, London NW1, UK
Hillside Laboratories, Route 1, Box 302, Clarkston, Washington, USA
Howson-Algraphy Ltd (Northern), Ring Road, Seacroft, Leeds 14, UK
Howson-Algraphy Ltd (Southern), Murray Road, Orpington, Kent, UK
Huck Co Inc, 1 Glenview Road, Montvale, New Jersey, USA
Hunter Associates Laboratory Inc, 9529 Lee Highway, Fairfax, Virginia,
    USA
Hygrodynamics Inc, 949 Selim Road, Silver Springs, Maryland, USA

Industrial Photo Equipment Ltd, Ipel House, 68 Paul Street, London EC2,
    UK
Inmont Corporation, Printing Ink Division, 67 W 44th St, New York,
    NY, USA
Instron Corporation, 2500 Washington St, Canton, Mass, USA
Instron Ltd, Materials Testing Instruments, Halifax Road, Cressex, High
    Wycombe, Bucks, UK

Appendix B

International Engineering Concessionaries Ltd, Wellington House,
Walton-on-Thames, Surrey, UK

Jacobsen, Van Den Berg & Co Ltd, Jacoberg House, Emerald Street,
London WC1, UK

K and N Laboratories, 1985 Anson Avenue, Melrose Park, Illinois, USA
Kent Engineering and Foundry Ltd, Maidstone, Kent, UK
Kollmorgen Corporation, Color Systems Division, 67 Mechanic St,
Attleboro, Mass, USA
Kollmorgen UK Ltd, Macbeth Division, 219 Kings Road, Reading,
Berks, UK

Lhomargy, 3 Boulevard de Bellevue, 91 Draveil, (Essonne) France
Liberty Engineering Co, Beloit, Wisconsin, USA
S R Littlejohn & Co Ltd, 16–24 Brewery Road, London N7, UK
Livingston Electronic Ltd, Livingston House, Greycaines Road, North
Watford, Herts, UK
Joyce Loebl & Co Ltd, Princes Way, Team Valley, Gateshead 11,
Co Durham, UK
Lorentzen & Wettres, Maskinaffar, Stockholm, Sweden
Lorilleux & Bolton Ltd, Eclipse Works, Ashley Road, Tottenham,
London N17, UK

Macbeth Corporation of Kollmorgen Corporation, PO Box 950, Newburgh,
NY, USA
Mander Kidd Ltd, P.O. Box 13, Old Heath Road, Wolverhampton, UK
Manufacturers Engineering and Equipment Corporation, 250 Titus Avenue,
Warrington, Pa, USA
Manufacturers Engineering & Equipment Corporation, 413a Brixton Road,
London SW9, UK
Marconi Instruments Ltd, Longacres, St Albans, Herts, UK
J B Marr & Co Ltd, Beacon Lodge, Strawberry Vale, Twickenham,
Middlesex, UK
Martin Sweets Co Inc, 3131 West Market Street, Louisville, Kentucky, USA
H E Messmer Ltd, 144c Offord Road, London N1, UK
Metallurgical Services Laboratories Ltd, Reliant Works, Betchworth,
Surrey, UK
Rudolf Meyer's Inc, 152–154 Brouwersgracht, Amsterdam, Holland
Mickle Laboratory Engineering Co, Mill Works, Gomshall, Nr Guildford,
Surrey, UK

Micro-Essential Laboratory Inc, 4224 Avenue H, Brooklyn, NY, USA
Moisture Register Co, 1510 W. Chestnut St, Alhambra, California, USA
Poul Murer, 12 Aagaardsveh, Vordingborg, Denmark

National Printing Ink Research Institute, Lehigh University, Bethlehem,
  Pennsylvania, USA
Neotec Instruments Inc, 640 Lofstrand Lane, Rockville, Maryland, USA
Newbold & Bulford Ltd, Enbeeco House, Roger Street, Gray's Inn Road,
  London WC1, UK
Norcross Corporation, 255 Newronville Avenue, Newton, Mass, USA
Nuclear Enterprises, Beenham Road, Reading, Berks, UK

Optica UK Ltd, Higham Lodge, Blackhorse Lane, Walthamstow,
  London E17, UK

Paper Facts and Figures, 10–16 Elm Street, London WC1, UK
B F Perkins, Standard International Corporation, P.O. Box 366, Chicopee,
  Mass, USA
Perkins-Elmer Corporation, Coleman Instruments Division, 42 Madison St,
  Maywood, Illinois, USA
Perkin Elmer Ltd, Beaconsfield, Bucks, UK
Phillips Engineering, Oldmixon Crescent, Oldmixon, Weston-Super-Mare,
  Somerset, UK
Phototronic Inc, 411 Cheltena Avenue, Jenkintown, Pa, USA
Photovolt Corporation, 1115 Broadway, New York, NY, USA
PIRA, Randalls Road, Leatherhead, Surrey, UK
Platemakers Educational & Research Institute, 2447 Western Avenue, Park
  Forest, Illinois, USA
Polyscience Corporation, 909 Pitner Avenue, Evanston, Illinois, USA
Precision Gage and Tool Co, 28 Volkenand Avenue, Dayton, Ohio
Pretema Ltd, Birmensdorf, Zurich, Switzerland
Promin Ltd, Bilborough, Yorkshire, UK
W G Pye & Co, P.O. Box 60, Cambridge, UK

Quantametric Devices Inc, 23 Jenison Avenue, Johnson City, NY, USA

Rank Precision Industries Ltd, Analytical Division, 31 Camden Road,
  London NW1, UK
Research Equipment Ltd, 64 Wellington Road, Hampton Hill, Middlesex,
  UK

Rex Gauge Co Inc, P.O. Box 46, Glenview, Illinois, USA
RK Chemical Co Ltd, South View Laboratories, Litlington, Royston, Herts, UK
Robbins Instrument Co Inc, 57 W 38th St, New York, NY, USA

Sangamo Controls Ltd, North Bersted, Bognor Regis, Sussex, UK
Sargent-Welch Scientific Co, 7300 N. Linder Avenue, Skokie, Illinois, USA
K E Saunders Agencies, P.O. Box 870, Port Washington, NY, USA
CZ Scientific Instruments Ltd, 93–97 New Cavendish Street, London W1, UK
Scott Tester Inc (Bendix Corporation), 101 Blackstone Street, Providence, Rhode Island, USA
Sheen Instruments Ltd, Sheendale Road, Richmond, Surrey, UK
Shirley Developments Ltd, P.O. Box 6, Wilmslow Road, Didsbury, Manchester, UK
Shore Instrument Co Inc, 90–35 Van Wyck Expressway, Jamaica, NY, USA
Soniscop Electronique, 100 Rue des Soldats, Berchem, Bruxelles 8, Belgium
Stanhope-Seta Ltd, Park Close, Englefield Green, Egham, Surrey, UK
Stouffer Graphic Arts Equipment Co, 311 N. Niles Avenue, South Bend, Indiana, USA

Taber Instruments Co, 455 Bryant St, N. Tonawanda, NY, USA
C F Taylor /Electronics) Ltd, Blackwater Station Estate, Camberley, Surrey, UK
Technical Section, British Paper and Board Makers' Association, Plough Place, Fetter Lane, London EC4, UK
Teledyne Inc, Automated Specialities Division, P.O. Box 888, Charlottesville, Virginia, USA
Tensometer Ltd, 77 Morland Road, Croydon, Surrey, UK
Testing Machines Inc, 400 Bayview Avenue, Amityville, Long Island, NY, USA
Testing Machines International of Canada Ltd, 6087 Sherbrooke St, W. Montreal, 261, Quebec, Canada
Thwing-Albert Instrument Co, 10960 Dutton Road, Philadelphia, Pa
Tinius Olsen Testing Machine Co, Easton Road, Willow Grove, Pa, USA
Tintometer, 6 Forest Drive, Jericho, Long Island, NY, USA
Tintometer Ltd, Waterloo Road, Salisbury, Wilts, UK
Tokyo Shivaura Electric Co Ltd, 1–6, 1-Chome, Uchisaiwai-Cho, Chiyoda-Ku, Tokyo, Japan
Toledo Scale Co, Toledo, Ohio, USA

Townson & Mercer Ltd, 101 Beddington Lane, Croydon, Surrey, UK
Turbojet Humidity and Temperature Control and Instrumentation Ltd,
   41 Foxley Lane, Purley, Croydon, Surrey, UK

Unicam Instruments Ltd, York Street, Cambridge, UK

Van der Korput, Baarn, Holland
Viscosel Corporation, P.O. Box 240, Stoughton, Mass, USA

H W Wallace & Co Ltd, St James Road, Croydon, Surrey, UK
William Webb Ltd, Perrymans Farm Road, Ilford, Barkingside, Essex, UK
Wennberg Apparater AB, Gubbangsvagen 113, Stockholm, Sweden
Wentworth Instruments Ltd, North Green, Datchet, Slough, Bucks, UK
Winstones Ltd, Park Works, Harefield, Middlesex, UK

Carl Zeiss, Inc, 444 Fifth Avenue, New York, NY, USA
Carl Zeiss, Oberkochen, West Germany

Appendix C

The following organisations have each published standard testing methods for printing materials. Lists of BS and TAPPI standards for testing paper and printing ink are set out in this appendix.

American Oil Chemists' Society (AOCS)
35 East Wacker Drive
Chicago, Illinois, USA

American Society for Testing and Materials (ASTM)
1916 Race Street
Philadelphia, Pennsylvania, USA

British Paper and Board Makers Association
Technical Section
Plough Place
Fetter Lane, London EC4, UK

British Standards Institution (BSI)
British Standards House
2 Park Street
London W1, UK

(Deutsche Industrial Normen-DIN)
Deutschen Normenausschuss (DNA)
Durggrafenstrasse 4–7
1 Berlin 30, Germany

Federal Test Method Standards (FTM)
Superintendent of Documents
Government Printing Office
Washington, DC, USA

Gravure Research Institute, Inc (GRI)
22 Manhasset Avenue, Manorhaven
Port Washington, New York, USA

National Printing Ink Research Institute (NPIRI)
Lehigh University
Bethlehem, Pennsylvania, USA

Packaging Institute (PI)
342 Madison Avenue
New York, NY, USA

Technical Association of the Pulp and Paper Industry (TAPPI)
360 Lexington Avenue
New York, NY, USA

United States of America Standards Institute (USA)
10 East 40th Street
New York, NY, USA

BRITISH STANDARDS INSTITUTION (BSI)

BS 2644:1955   Sizing properties of paper, method of testing the degree of water resistance.

BS 2699:1956   Method for determining the absorbency of blotting paper: Determination of the ink absorbency time.

BS 2916:1957   Absorbency test for bibulous paper.

BS 2922:1958   Methods for testing the strength of wet paper.

BS 2924:1968   Determination of the pH value, conductivity and chloride and sulphate contents of aqueous extracts of paper and board.

BS 2925:1958   Methods for determining air permeability and air resistance of paper.

BS 2987:1958   Notes on the application of statistics to paper testing.

BS 3137:1959   Method for determining the bursting strength of paper.

BS 3177:1959   Permeability to water vapour of flexible sheet materials used for packaging.

BS 3430:1968   Sampling of paper and board for testing.

BS 3431:1961   Conditioning of paper and board test samples.

BS 3432:1971   Determination of the grammage (basis weight) of paper and board.

BS 3433:1961   Sampling and testing of paper for moisture content.

BS 3631:1963   Method for the determination of the ash content of paper.

BS 3748:1964   Determination of stiffness of board.

BS 3755:1964 Assessment of odour from packaging materials used
for foodstuffs.
BS 3983:1968 Determination of the thickness and bulk of paper.
BS 4415:1969 Determination of tensile strength of paper and board.
BS 4419:1969 Recommendations for measurement of folding
endurance of paper.
BS 4420:1969 Recommendations for determination of Bendtsen
roughness of paper and board.
BS 4432:1969 Methods for determining optical properties of pulp, paper
and board.
Part 1–Determination of diffuse ISO reflectance
factor of pulp, paper and board.
Part 2–Measurement of diffuse ISO brightness
(blue reflectance factor) of paper and board.
Part 3–Measurement of ISO opacity
(paperbacking) of paper.
BS 4468:1969 Method for the determination of the internal
tearing resistance of paper.
BS 4497:1969 Recommendations for the detection and estimation
of nitrogenous treating agents in paper.
BS 4574:1970 Recommendations for the determination of the
ink absorbency of paper and board.
BS 4616:1970 Recommendations for the determination of
compressibility of paper and board (Bendtsen method).
BS 4685:1971 Determination of wax content of waxed paper and board.
BS 4686:1971 Determination of flat crush resistance of
corrugated fibreboard.

BS 1480:1949 Four and three colour letterpress inks.
BS 2650:1955 Four colour offset lithographic inks.
BS 188:1957 Viscosity.
BS 1733:1955 Flow Cups.
BS 3020:1959 Improved inks for three and four colour letterpress
printing.
BS 3110:1959 Rub-resistance of print.
BS 1006:1961 Fastness to light.
BS 4160:1967 Inks for letterpress three and four colour printing.
BS 4321:1969 Methods of test for printing inks.
BS 4666:1971 Inks for offset three- and four-colour printing.

TECHNICAL ASSOCIATION OF THE PULP AND PAPER INDUSTRY (TAPPI)

T 400 os – 70	Sampling of Paper and Paperboard
T 401 m – 60	Fibre Analysis of Paper and Paperboard
T 402 os – 70	Conditioning Paper and Paperboard for Testing
T 403 ts – 63	Bursting Strength of Paper
T 404 ts – 66	Tensile Breaking Strength of Paper and Paperboard
T 405 ts – 63	Wax in Paper
T 406 m – 60	Reducible Sulfur in Paper
T 408 os – 61	Rosin in Paper and Paperboard
T 409 cs – 61	Machine Direction of Paper
T 410 os – 68	Weight per Unit Area (Basis Weight or Substance) of Paper and Paperboard
T 411 os – 68	Thickness (Caliper) of Paper and Paperboard
T 412 os – 69	Moisture in Paper and Paperboard
T 413 ts – 66	Ash in Paper
T 414 ts – 65	Internal Tearing Resistance of Paper
T 415 os – 68	Casein and Soya Protein in Paper (Qualitative)
T 417 os – 68	Proteinaceous Nitrogenous Materials in Paper (Qualitative)
T 418 os – 61	Organic Nitrogen in Paper
T 419 su – 70	Starch in Paper
T 421 os – 61	Qualitative Analysis of Mineral Filler and Mineral Coating of Paper
T 422 su – 67	Quantitative Analysis of Mineral Filler and Mineral Coating of Paper
T 423 su – 68	Folding Endurance of Paper (Schopper Tester)
T 425 m – 60	Opacity of Paper
T 427 su – 71	Saturating Properties of Roofing Felt
T 428 su – 67	Hot Water Extractable Acidity or Alkalinity of Paper
T 429 os – 69	Alpha-Cellulose in Paper
T 430 su – 68	Copper Number of Paper and Paperboard
T 431 ts – 65	Ink Absorbency of Blotting Paper
T 432 ts – 64	Water Absorption of Bibulous Paper
T 433 m – 44	Water Resistance of Paper and Paperboard (Dry-Indicator Method)
T 434 os – 68	Acid-Soluble Iron in Paper
T 435 su – 68	Hydrogen Ion Concentration (pH) of Paper Extracts— Hot Extraction Method
T 436 ts – 64	Arsenic in Paper
T 437 ts – 63	Dirt in Paper and Paperboard

Appendix C

T 438 ts – 65 Zinc and Cadmium in Paper
T 439 m – 60 Titanium Pigments in Paper
T 440 su – 70 Alkali-Staining Number of Paper
T 441 os – 69 Water Absorptiveness of Nonbibulous Paper and
          Paperboard (Cobb Test)
T 442 m – 47 Spectral Reflectivity and Color of Paper
T 444 os – 68 Silver Tarnishing by Paper
T 445 sm – 57 Identification of Specks and Spots in Paper
T 448 su – 71 Water Vapor Permeability of Paper and Paperboard
T 449 os – 64 Bacteriological Examination of Paper and Paperboard
T 451 m – 60 Rigidity, Stiffness and Softness of Paper
T 452 m – 58 Brightness of Paper and Paperboard
T 453 su – 70 Effect of Heat on Folding Endurance (Relative Stability of
          Paper)
T 454 ts – 66 Turpentine Test for Grease Resistance of Paper
T 455 os – 68 Identification of Wire Side of Paper
T 456 os – 68 Wet Tensile Breaking Strength of Paper and Paperboard
T 457 m – 46 Stretch of Paper and Paperboard
T 458 os – 70 Surface Wettability of Paper (Angle-of-Contact Method)
T 459 su – 65 Wax Pick Test of Paper
T 460 os – 68 Air Resistance of Paper
T 461 os – 68 Flame Resistance of Treated Paper and Paperboard
T 462 su – 71 Printing-Ink Receptivity of Paper (Castor-Oil Test)
T 463 su – 70 Adhesiveness of Gummed Paper Tape
T 464 su – 71 Water Vapor Permeability of Sheet Materials at High
          Temperature and Humidity
T 465 su – 71 Creasing of Paper for Water Vapor Permeability Tests
T 466 m – 52 Degree of Curl and Sizing of Paper
T 467 m – 48 Paraffin Wax Absorptiveness of Paper
T 468 m – 60 Water-Soluble Sulfates and Chlorides in Paper and
          Paperboard
T 470 os – 66 Edge Tearing Resistance of Paper
T 471 m – 47 Testing Analytical Filter Papers
T 472 su – 68 Compression Resistance of Paperboard (Ring Crush Test)
T 475 su – 71 Bleeding Resistance of Asphalted Paper
T 476 ts – 63 Abrasion Loss of Packaging Materials
T 477 m – 47 Blocking Resistance of Paper and Flexible Materials
T 479 su – 71 Smoothness of Printing Paper (Bekk)
T 480 ts – 65 Specular Gloss of Paper and Paperboard at 75°
T 481 sm – 60 Fiber Orientation in Paper (Zero-Span Tensile Strength)

T 482 m – 52   Water Vapor Permeability of Sheet Materials at 0°F.
T 483 sm – 53   Odor of Packaging Materials
T 484 m – 58   Moisture in Paper by Toluene Distillation
T 486 su – 69   Blood Resistance of Butchers' Wrapping Paper
T 487 m – 54   Fungus Resistance of Paper and Paperboard
T 488 ts – 65   Microscopical Identification of Fillers in Paper
T 489 os – 70   Stiffness of Paperboard
T 490 sm – 58   Smoothness of Paper under 3 psi Clamping Pressure
T 491 su – 63   Water Immersion Test of Paperboard
T 492 sm – 60   Water Absorption of Paperboard
T 493 su – 68   Identification and Determination of Melamine Resin in Paper
T 494 os – 70   Tensile Breaking Properties of Paper & Paperboard
T 495 ts – 66   Bending Number of Paperboard
T 496 su – 64   Cross Directional Internal Resistance of Paperboard
T 497 os – 69   Surface Wax on Waxed Paper or Paperboard
T 498 su – 66   Softness of Sanitary Tissues
T 499 su – 64   Surface Strength of Paper (IGT Tester)
T 500 ts – 65   Book Bulk and Bulking Number of Paper
T 501 su – 67   Wetting Shipping Sack Paper for Testing
T 502 su – 67   Equilibrium Relative Humidity of Paper and Paperboard
T 503 su – 67   Coefficient of Static Friction of Shipping Sack Papers
T 504 su – 67   Quantitative Determination of Glue in Paper
                      (Hydroxyproline Assay)
T 505 su – 67   Qualitative Identification of Glue in Paper
T 506 su – 68   Internal Bond Strength of Paper and Paperboard
                      (Z-directional Tensile Test)
T 507 su – 68   Grease Resistance of Flexible Packaging Materials
T 508 su – 68   Illuminants (Wave Lengths 400–700 nm only) for Visual
                      Grading and Color Matching of Paper
T 509 su - 68   Hydrogen Ion Concentration (pH) of Paper Extracts—
                      Cold Extraction Method
T 510 su – 69   Water Resistance of Adhesive Bond in Laminated Paper
                      and Paperboard
T 511 su – 69   Folding Endurance of Paper (MIT Tester)
T 512 su – 69   Creasing of Flexible Packaging Material Specimens for
                      Testing
T 513 su – 69   Water Repellency of Paper and Boards
T 514 su – 69   Surface Strength of Coated Paperboard
T 515 su – 70   Visual Grading and Color Matching of Paper with an
                      Ultraviolet-Containing Daylight Illuminant

Appendix C

Appendix D

Metric units are based on a decimal system which is easy to use because of the logical inter-relationships and simplicity of calculation. SI units (Systeme International d'Unites) provide a modern form of the metric system agreed at an international conference in 1960. From the six basic or primary units, a series of supplementary or derived units may be obtained. The multiples and submultiples of the units, expressed by prefixes, are the same regardless of the units to which they are applied.

Primary units

Quantity	*Unit*	*Symbol*
length	metre	m
mass	kilogramme	kg
time	second	s
electric current	ampere	A
temperature	kelvin	K
luminous intensity	candela	cd

Some supplementary and derived units

Quantity	*Unit*	*Symbol or abbreviation*
Area	square metre	m^2
Volume	cubic metre	m^3
Velocity	metre per second	m/s
Acceleration	metre per second squared	m/s^2
Frequency	hertz	Hz (s^{-1})
Mass Density	kilogramme per cubic metre	kg/m^3

278

Appendix D

Quantity	*Unit*	*Symbol or abbreviation*
Momentum	kilogramme metre per second	kg m/s
Angular momentum	kilogramme metre squared per second	kg m²/s
Moment of inertia	kilogramme metre squared	kg m²
Force	newton	N (kg m/s²)
Moment of force	newton metre	N m
Pressure, stress	pascal	Pa (N/m²)
Viscosity: kinematic	metre squared per second	m²/s
dynamic	newton second per metre squared	Pas (Ns/m²)
Surface tension	newton per metre	N/m
Work, energy, quantity of heat	joule	J (Nm)
Power, heat flow rate	watt	W (J/s)
Impact strength	joule per square metre	J/m²
Temperature (customary unit)	degree Celsius	°C
Temperature interval	kelvin, degree Celsius	K,°C
Thermal coefficient of linear expansion	reciprocal degree Celsius or reciprocal kelvin	°C⁻¹ K⁻¹
Thermal conductivity	watt per metre degree C	W/m °C
Quantity of electricity, electric charge	coulomb	C (As)
Electric tension, potential difference electromotive force	volt	V (W/A)
Electric field strength	volt per metre	V/m
Electric resistance	ohm	Ω (V/A)
Electric capacitance	farad	F (AS/V)
Magnetic flux	weber	Wb (Vs)
Inductance	henry	H (Vs/A)
Magnetomotive force	ampere	A
Luminous flux	lumen	lm (cd sr)
Luminance	candela per square metre	cd/m²
Illumination	lux	lx (lm/m²)

Prefixes for multiples and sub-multiples of SI units

Prefix	Symbol		Factor by which the unit is multiplied
tera	T	10^{12} =	1 000 000 000 000
giga	G	10^9 =	1 000 000 000
mega	M	10^6 =	1 000 000
kilo	k	10^3 =	1 000
hecto*	h	10^2 =	100
deca*	da	10^1 =	10
deci*	d	10^{-1} =	1·0
centi*	c	10^{-2} =	0·01
milli	m	10^{-3} =	0·001
micro	μ	10^{-6} =	0·000 001
nano	n	10^{-9} =	0·000 000 001
pico	p	10^{-12} =	0·000 000 000 001
femto	f	10^{-15} =	0·000 000 000 000 001
atto	a	10^{-18} =	0·000 000 000 000 000 001

* The prefixes marked with an asterisk should be limited as far as possible for use where other recommended prefixes are inconvenient.

Some useful conversions to SI units

Length	1 metre (m) = 39·370in
	1in = 25·4mm
Mass	1 kilogramme (kg) = 2·2046lb
	1lb = 453·59g
Area	1 square metre (m²) = 1550·0in²
Volume	1 cubic metre (m³) = 35·314ft³
	1 litre (1) = 0·219 98 Imp. gal
Force	1 newton (N) = 0·224 81 lbf
Work	1 joule (J) = 0·757 56ft lbf
Pressure	1 newton per metre squared (N/m²) = 0·145 04 × 10^{-3}lbf/in²
Dynamic viscosity	1 newton second per metre squared (Ns/m²) = 10 poise (p)
Surface tension	1 newton per metre (N/m) = 1000 dyne/cm

Appendix D

SI Units for use in the testing of paper and ink

Property	Unit used in the past	SI Unit	Symbol	Conversion factor
Grammage (substance or basis weight)	gramme per square metre	gramme per square metre	g/m^2	
Thickness	mil	micrometre (micron)	μm	$1\text{mil} = 0{\cdot}001\text{in}$ $= 25{\cdot}4\mu m$
Static tensile strength	pound per inch width	newton per 15mm width	N/15mm	$1\text{lbf/in} = 2{\cdot}627\,\text{N/15mm}$
Dynamic tensile strength	centimetre kilogramme per inch width	millijoule per 15mm width	mJ/15mm	$1\text{cm kgf/in} = 57{\cdot}91\text{mJ/15mm}$
Bursting strength	pound per square inch	kilopascal	kPa	$1\text{lbf/in}^2 = 6{\cdot}895\text{kPa}$
Tearing strength	gramme	millinewton	mN	$1\text{gf} = 9{\cdot}807\text{mN}$
Stiffness	gramme	millinewton	mN	$1\text{gf} = 9{\cdot}807\text{mN}$
IGT picking velocity	centimetre per second	millimetre per second	mm/s	$1\text{cm/s} = 10\text{mm/s}$
Dynamic viscosity	poise	newton second per metre squared, or pascal second	Ns/m^2 Pa s	$1Ns/m^2 = 10p$ $1Pa\,s = 10p$

Appendix E

International Paper Sizes

The application of metrication to paper sizes is much more than the conversion of existing paper dimensions into metric equivalents. The printing and paper industries have decided that as far as possible the sizes recommended by the International Organisation for Standardisation (ISO) should be brought into use. These standards allow for the variety needed yet will eliminate unnecessary sizes and simplify manufacture, storage, purchasing and marketing.

The Principle of ISO Sizes

The system of ISO sizes is based on three series of trimmed sizes. All of these have the same proportion and are designated A, B and C. The most widely used is the A series for stationery and general leaflet printing. The B series is intended primarily for larger printed items such as posters and wall charts. The C series, in conjunction with some of the B sizes, is used for envelopes.

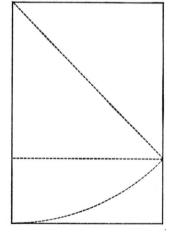

All sheets in the three series of sizes are of the same shape, that of a rectangle whose sides are in the ratio of 1 to the square root of 2. This shape, where a diagonal of a square becomes the long side of a rectangle based on a square (see illustration) has been known to architects and designers throughout the ages as the 'golden square' and is recognised as being a perfectly balanced rectangle. A particular sheet size in each of the series is obtained by dividing, parallel to the shorter side, the size immediately above it into two equal parts.

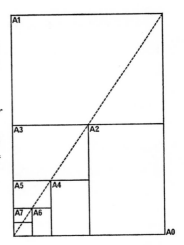

The A series of sizes is based on the A0 sheet which has an area of one square metre. The A1 size equals half of A0, similarly A2 is half of A1, A3 is half of A2 and so on (see illustration). The size of one square metre allows the direct use of grammes per square metre for designating the substance of the paper. The constant ratio between the long and short sides of the sub-divisions of the A0 sheet means that any drawings, artwork, diagrams or type setting prepared for one size of the series will be in proportion to any other ISO paper size.

Trimmed and Untrimmed sizes

During the investigations by the ISO a fundamental difference of practice in regard to the whole basis of arriving at standard sizes of paper was highlighted. British practice was to use the untrimmed size as the basis from which the trimmed sizes were obtained by removing a variable amount of 'trim'. In many countries the trimmed size was fixed regardless of the untrimmed size from which it was obtained. To clarify the situation the following definitions were used.

Untrimmed Size

Dimensions of a sheet of paper, untrimmed and not specially squared, sufficiently large to allow a trimmed size to be obtained from it as required.

Trimmed Size

The final dimensions of a sheet of paper.

To avoid possible confusion among papermakers, who may interpret 'untrimmed' and 'trimmed' as meaning 'not guillotine trimmed' and 'guillotine trimmed' respectively the term 'stock sizes' was introduced and applied to the untrimmed sizes. This implies that the paper or board may be guillotine trimmed or not but that it will normally require further trimming by the printer before it is ready for use by the final user.

The prefix R signifies untrimmed sizes from which the A and B sizes may be obtained. The prefix S R signifies sizes that may be trimmed to bleed.

ISO TRIMMED AND UNTRIMMED STOCK SIZES AND THEIR INCH CONVERSIONS

'ISO-A' series of trimmed sizes

Designation	Size mm	Inch Conversion
A0	841 × 1189	33·11 × 46·81
A1	594 × 841	23·39 × 33·11
A2	420 × 594	16·54 × 23·39
A3	297 × 420	11·69 × 16·54
A4	210 × 297	8·27 × 11·69
A5	148 × 210	5·83 × 8·27
A6	105 × 148	4·13 × 5·83
A7	74 × 105	2·91 × 4·13
A8	52 × 74	2·05 × 2·91
A9	37 × 52	1·46 × 2·05
A10	26 × 37	1·02 × 1·46

'ISO-B' series of trimmed sizes

B0	1000 × 1414	39·37 × 55·67
B1	707 × 1000	27·83 × 39·37
B2	500 × 707	19·68 × 27·83
B3	353 × 500	13·90 × 19·68
B4	250 × 353	9·84 × 13·90
B5	176 × 250	6·93 × 9·84
B6	125 × 176	4·92 × 6·93
B7	88 × 125	3·46 × 4·92
B8	62 × 88	2·44 × 3·46
B9	44 × 62	1·73 × 2·44
B10	31 × 44	1·22 × 1·73

Appendix E

RA series of untrimmed stock sizes

Reels		Width mm	Inch Conversion
		430	16·93
		610	24·02
		860	33·86
		1220	48·03

Sheets	Designation	Size mm	Inch Conversion
	RAo	860 × 1220	33·86 × 48·03
	RA1	610 × 860	24·02 × 33·86
	RA2	430 × 610	16·93 × 24·02

SRA series of untrimmed stock sizes

Reels		Width mm	Inch Conversion
		450	17·72
		640	25·20
		900	35·43
		1280	50·39

Sheets	Designation	Size mm	Inch Conversion
	SRAo	900 × 1280	35·43 × 50·39
	SRA1	640 × 900	25·20 × 35·43
	SRA2	450 × 640	17·72 × 25·20

Supplementary sizes

In addition to the ISO range of sizes, additional sizes of Printing and Writing papers have been introduced at the request of the British Federation of Master Printers. These are intended to replace the traditional Crown range of sizes of 20in × 30in and its multiples. These so called metric Crown sizes are not exact conversions of the Imperial sizes since it was considered preferable to standardise on a single range of paper sizes for both book printing and general printing. The metric Crown range are as follows:

Quad Crown	768 × 1008mm
Double Crown	504 × 768mm

Appendix F

Grammage or Basis weight (Substance)

Standardisation will also be introduced in respect of substance. The ream will be standardised at 500 sheets. Substance will be expressed as the weight in grammes of one square metre of one sheet of paper. Substance will in future be known as 'Grammage' or 'Basis Weight'. The Basis Weight of a paper will, therefore, be constant irrespective of the size and number of sheets. The traditional method of expressing substance as pounds per ream will be discontinued.

For the designation of Basis Weight, British and International standards specify the correct abbreviation of grammes per square metre to be g/m^2. Where typing facilities are available to use this symbol without extra cost or inconvenience it will be used and where the symbol can be readily printed it can also be used. In other circumstances the abbreviation gsm, in lower case and without any punctuation, will be recognised as the abbreviation for grammes per square metre.

The R20 and R40 ranges

The R series (R10, R20, R40, etc) of numbers – named after Charles Renard, the French engineer who first proposed their use – are now widely used in industry generally. The number indicates the particular root of 10 on which the series is based. Thus the R20 series ($20\sqrt{10}$ or $1\cdot12$) is one in which each step increases by approximately 12% over the preceding step ie, is multiplied by $1\cdot12$ and 'rounded off'.

A new standard range of Basis Weights related to the R20 series of preferred numbers will be introduced for some stock papers. The series is a geometrical progression and provides a greater choice of Basis Weights at the lighter end of the range which makes it very suitable for commercial requirements.

The R20 series is intended mainly for stock papers but it will be possible to choose any Basis Weight from it for 'making' orders. It may sometimes be necessary to select a Basis Weight from the R40 range, which is an intermediate series, each number being approximately midway between each pair of R20 series numbers.

THE R20 AND R40 RANGES OF BASIS WEIGHTS

in grammes per square metre

R20

20·0	40·0	80·0	160·0	315·0
22·4	45·0	90·0	180·0	355·0
25·0	50·0	100·0	200·0	400·0
28·0	56·0	112·0	224·0	
31·5	63·0	125·0	250·0	
35·5	71·0	140·0	280·0	

R40

21·2	42·5	85·0	170·0	335·0
23·6	47·5	95·0	190·0	375·0
26·5	53·0	106·0	212·0	
30·0	60·0	118·0	236·0	
33·5	67·0	132·0	265·0	
37·5	75·0	150·0	300·0	

Appendix G

TRADITIONAL PAPER SIZES IN BRITAIN AND THE UNITED STATES

(a) Traditional British Sizes

Name of size	Size in inches	Size in millimetres	Factor for converting lb/ream (500 sheets) to g/m²
Foolscap	17 × 13½	432 × 343	6·13
Large Post	21 × 16½	533 × 419	4·06
Demy	22½ × 17½	572 × 445	3·57
Medium	23 × 18	584 × 457	3·40
Double Crown	20 × 30	508 × 762	2·34

(b) Basic USA Sizes

General Group	Basic size in inches	Area in sq. in.	Size in millimetres
Bond, ledger, manifold railroad manila, writing	17 × 22	374	432 × 559
Blotting	19 × 24	456	483 × 610
Box cover, cover	20 × 26	520	508 × 660
Manuscript cover	18 × 31	558	457 × 787
Blanks, tough check	22 × 28	616	559 × 711
Mill bristol, postcard, tag, wedding bristol	22½ × 28½	641	572 × 724
Index bristol	25½ × 30½	778	648 × 775
Mill bristol	22½ × 35	786	572 × 889
Box cover, glassine, hanging, newsprint, poster, tag, tissues, waxing tissues, wrapping, wrapping tissues	24 × 36	864	610 × 914
Bible, book, box cover, gummed, offset	25 × 38	950	635 × 965

Index

Index

Index

Index

Index